IFSTA 107
SIXTH EDITION

FIRE VENTILATION PRACTICES

VALIDATED BY
THE INTERNATIONAL FIRE SERVICE TRAINING ASSOCIATION

PUBLISHED BY
**FIRE PROTECTION PUBLICATIONS
OKLAHOMA STATE UNIVERSITY**

Cover Photo

University of Illinois Fire Service Institute students practice trenching and other ventilation techniques on an old horse barn. Photo by John C. Dixon, Champaign-Urbana (Ill.) *News-Gazette.*

Dedication

This manual is dedicated to the members of that unselfish organization of men and women who hold devotion to duty above personal risk, who count sincerity of service above personal comfort and convenience, who strive unceasingly to find better ways of protecting the lives, homes and property of their fellow citizens from the ravages of fire and other disasters . . . The firefighters of All Nations.

Dear Firefighter:

The International Fire Service Training Association (IFSTA) is a nonprofit organization that exists for the sole purpose of serving firefighters. IFSTA is a member of the Joint Council of National Fire Organizations, National Fire Protection Association and International Society of Fire Service Instructors. If you need help in locating additional information concerning the organization, training materials, or manual orders, please write or call.

Without you we do not exist

Write:
Customer Services
Fire Protection Publications
IFSTA Headquarters
Oklahoma State University
Stillwater, Oklahoma 74078

Call:
(405) 624-5723

FIRST PRINTING - December, 1980
SECOND PRINTING - September, 1981

ISBN 0-87939-039-5
Library of Congress 80-84149
Sixth Edition
Printed in the United States of America

THE INTERNATIONAL FIRE SERVICE TRAINING ASSOCIATION

The International Fire Service Training Association is an educational alliance organized to develop training material for the fire service. Each year, validation committees meet at a workshop conference, the objectives of which are to:

develop training material for publication;
validate training material for publication;
check proposed rough drafts for errors;
add new techniques and developments;
delete obsolete and outmoded methods; and
upgrade the fire service through training.

This training association was formed in November 1934, when the Western Actuarial Bureau sponsored a conference in Kansas City, Missouri, to determine how all agencies that were interested in publishing fire service training material could coordinate their efforts. Four states were represented at this conference and it was decided that since the representatives from Oklahoma had done some pioneering in fire training manual development other interested states should join forces with them. This merger made it possible to develop nationally recognized training material broader in scope than material published by any individual state agency. This merger further made possible a reduction in publication costs since it enabled each state to benefit from the economy of relatively large printing orders. These savings would not be possible if each individual state developed and published its own training material.

From the original four states, the adoption list has grown to almost all of the United States; all Canadian provinces; Bermuda; Queensland Australia; the International Civil Aviation Organization Training Centre, Beirut, Lebanon; the Department of National Defence of Canada; the Department of the Army of the United States; the Department of the Navy of the United States; the United States Air Force; the United States Bureau of Indian Affairs; the United States General Services Administration, and the National Aeronautics and Space Administration (NASA).

Every July, IFSTA's validation conference is held in Stillwater, Oklahoma. This conference is an invitational conference, bringing together persons from several related and allied fields:

key fire department executives and drillmasters;
educators from colleges and universities;
representatives of government agencies;
representatives of firefighter organizations; and
engineers from the fire insurance, equipment,
and apparatus industries.

The committees on which these persons serve validate the material contained in IFSTA manuals as up-to-date and correct for the fire service.

The publications of the International Fire Service Training Association are compatible with the National Fire Protection Association's Standard No. 1001, *Fire Fighter Professional Qualifications,* and the International Association of Fire Fighters/International Association of Fire Chiefs *National Apprenticeship and Training Standards for the Fire Fighter*. The standards are an effort to attain professional status through progressive training. The NFPA and IAFF/IAFC Standards were prepared in cooperation with the Joint Council of National Fire Service Organizations, of which IFSTA is a member.

To formally adopt IFSTA training material, the firefighters should first request it through their training agencies. Adopting does not obligate any sponsoring agency. It adds prestige to any training program and this, in turn, adds prestige to IFSTA publications.

PREFACE

This is the sixth edition of IFSTA 107. There have been some substantial changes, reflecting new techniques of building construction and their effects on the firefighter's ventilation procedures. There are many new illustrations and photographs added to this edition. Before this edition was validated by the IFSTA membership, many hours were spent considering what information to retain and how much should be changed. The editorial staff extends its gratitude to the validating committee for its invaluable aid:

Chairman
Paul Boecker
Fire Chief
Lisle-Woodridge Fire District
Lisle, Illinois

Secretary
Richard Giles
Director of Safety
Oklahoma State University
Stillwater, Oklahoma

Bill Darin
Deputy Chief, Fire Marshal
Edmond Fire Department
Edmond, Oklahoma

Joseph Donovan
Chief of Fire Training
Massachusetts Firefighting Academy
Sudbury, Massachusetts

Donald McKay
Fire Instructor
Minnesota Department of Education
St. Paul, Minnesota

Kenneth Mitten
Deputy Chief
Merced City Fire Department
Merced, California

Larry Northcutt
Training Officer
Norman Fire Department
Norman, Oklahoma

W. W. Smith
Specialist
Alabama State Department of Education
University, Alabama

Kenneth Stanton
Region Four Executive Coordinator
National Automatic Sprinkler and
Fire Control Association, Inc.
Indianapolis, Indiana

Max Thomas
Director
Fire Service Training
University of Kansas
Lawrence, Kansas

Our gratitude is also extended to other persons who served on the committee and offered their advice and assistance:

Thomas Campbell
E. F. Kaughman
Robert M. Porter
Jim Straseske
Thomas Woods
Earl Woolrich
Floyd Yokum

Much of the information contained in IFSTA manuals would be impossible to obtain and print if it were not for the openhanded assistance of various individuals, organizations, and private companies. We acknowledge our debt to the following, who kindly provided illustrations or gave permission to use copyrighted material:

Angus Fire Armour
Controlled Airstreams, Inc.
Lisle-Woodridge Fire District
Maxax Industries, Inc.
Nancy Engebretson
Super Vacuum Manufacturing Co., Inc.
Univ. of Illinois Fire Service Institute
Wilcox Silent Nozzleman

We must also acknowledge our debt to the following persons, without whose expertness and dedication this manual could not have its final form: Charles W. Orton, publications editor; Don Davis, production coordinator; Ann Moffat, artist; Janice Wiles and Karen Murphy, phototypesetter operators; and Carol Smith, proofreader.

Gene P. Carlson
Editor

CONTENTS

LIST OF TABLES

INTRODUCTION

Ventilation is the systematic removal of heated air, smoke, and gases from a structure, followed by the replacement of these products of combustion with a supply of cooler air, which facilitates other fire fighting priorities.

The importance of ventilation becomes obvious. It increases visibility for quicker location of the seat of the fire. It decreases the danger to trapped occupants by channeling away hot, toxic gases. It reduces the chance of flashover or backdraft. Unfortunately, ventilation may be misunderstood by the public because it requires doing limited damage to a building; but it results in a much larger reduction in damage.

Ventilation must be planned

Proper ventilation cannot be accomplished haphazardly or independently of a well-planned fire attack. Certain technical principles are involved. A firefighter cannot rely solely upon knowledge gained from practical experience, because no two fires are alike. Instead, the firefighter should begin with a basic study of ventilation theory and progress to the techniques of making the necessary openings in fire buildings for removal of accumulated combustion products.

Modern technology requires new emphasis on ventilation. Consider a single item: plastics. Since the middle of the twentieth century, the use of plastic materials has had phenomenal growth. These materials perform so vital a function that a new industry has been created, taking its place beside such basic industries as those using wood, metal, and textiles. The plastics industry is expected to surpass the steel industry in pounds produced per year early in the 1980s. As a result, the fuel load in all occupancies will be increased and one can also expect increases in the amounts of products of combustion produced during fires. Prompt and proper ventilation for quick and efficient saving of lives, suppression of fire, and reduction of damage becomes more important every day.

Indeed, a knowledge of ventilation is of paramount importance, second only to the application of an extinguishing agent. A fire officer with an understanding of what is taking place in a fire building and what effect certain optional actions will produce is much better prepared to assume responsibility in dealing with ventilation.

Understanding of ventilation necessary

In order to function capably as a team member, a firefighter needs to understand the characteristics and behavior of those elements and situations with which the firefighter must contend. It is not sufficient that ventilation be understood only by the senior fire officer in charge of a fire. There are numerous instances during every involved fire for which each firefighter must arrive at decisions and operate independently within the orders passed down from the senior officer. It is at this point that the value of this training "pays off." If the firefighter has actively participated in a well-balanced, well-organized training program, which includes all aspects of the profession, the firefighter will select the choice of alternatives that experience and training have taught is proper for the situation.

Ventilation aids other objectives

Major objectives of a fire department are to reach the scene of the fire as quickly as possible, rescue trapped victims, locate the fire, and apply suitable extinguishing agents with a minimum of fire, water, smoke, and heat damage. Ventilation during fire fighting is definitely an aid to the fulfillment of these objectives. To effectively and adequately achieve some of these objectives, it often becomes necessary to safely get inside a structure. This problem emphasizes one reason why all fire companies should be equipped with adequate respiratory protection for each firefighter. Although fire ventilation provides better breathing conditions within a structure, it does not remove all hazards and dangerous gases. The application of water fog into a heavily charged, heated area may collect carbon particles from the smoke and disperse smoke and gases from the area, but it does not eliminate the need for respiratory protection.

Fire ventilation is conceded to be vitally essential in fire control, yet there are other factors and conditions that also influence the success or failure of an operation. Failure to provide for all conditions as they arise may jeopardize an entire operation. However, when proper ventilation is accomplished to aid fire control there are certain advantages that may be obtained from its application.

Purpose and Scope

The purpose of this manual is to present the principles and practices of ventilation in a manner that will provide a basis for training and enable the discerning student to effectively contribute to reducing the loss of life and property by fire.

The scope of this manual is intended to include factors related to products of combustion, elements and situations that influence the ventilation process, and methods and procedures of ventilating.

1 FIRE BEHAVIOR RELATED TO VENTILATION

NFPA STANDARD 1001

Fire Fighter I

3-10 Ventilation

3-10.1 The fire fighter shall define the principles of ventilation, and identify the advantages and effects of ventilation.

3-10.2 The fire fighter shall identify the dangers present, and precautions to be taken in performing ventilation.

3-10.6 The fire fighter shall define the theory of a "back draft explosion."

PHASES OF A FIRE

Fires may start at any time of the day or night if the hazard exists. If the fire happens when the area is occupied, the chances are that it will be discovered and controlled in the beginning phase. But if it occurs when the building is closed and deserted, the fire may go undetected until it has gained major proportions. The condition of a fire in a closed building is one of chief importance for ventilation.

When fire is confined in a building or room, a situation develops that requires carefully calculated and executed ventilation procedures if further damage is to be prevented and danger reduced. This type of fire can be best understood by an investigation of its three progressive stages.

A firefighter may be confronted by one or all of the following phases of fire at any time; therefore, a working knowledge of these phases is important for understanding of ventilation procedures.

Incipient or Beginning Phase

In the first phase, the oxygen content in the air has not been significantly reduced and the fire is producing water vapor (H_2O), carbon dioxide (CO_2), perhaps a small quantity of sulfur dioxide (SO_2), carbon monoxide (CO), and other gases. Some heat is being generated, and the amount will increase with the progress of the fire. The fire may be producing a flame temperature well above 1,000°F (537°C), yet the temperature in the room at this stage may be only slightly increased. (Figure 1.1.)

Figure 1.1

INCIPIENT PHASE

- Slightly Over 100°F (37.8°C) In The Room
- Rising Hot Gases
- Room Air Approximately 20% Oxygen

Free-Burning Phase

The second phase of burning encompasses all of the free-burning activities of the fire. During this phase, oxygen-rich air is drawn into the flame as convection (the rise of heated gases) carries the heat to the uppermost regions of the confined area. The heated gases spread out laterally from the top downward, forcing the cooler air to seek lower levels, and eventually igniting all the combustible material in the upper levels of the room (Figure 1.2). This heated air is one of the reasons that firefighters are taught to keep low and use protective breathing equipment. One breath of this superheated air can sear the lungs. At this point, the temperature in the upper regions can exceed 1,300°F (700°C). As the fire progresses through the latter stages of this phase, it continues to consume the free oxygen until it reaches the point where there is insufficient oxygen to react with the fuel. The fire is then reduced to the smoldering phase and needs only a supply of oxygen to burn rapidly or explode.

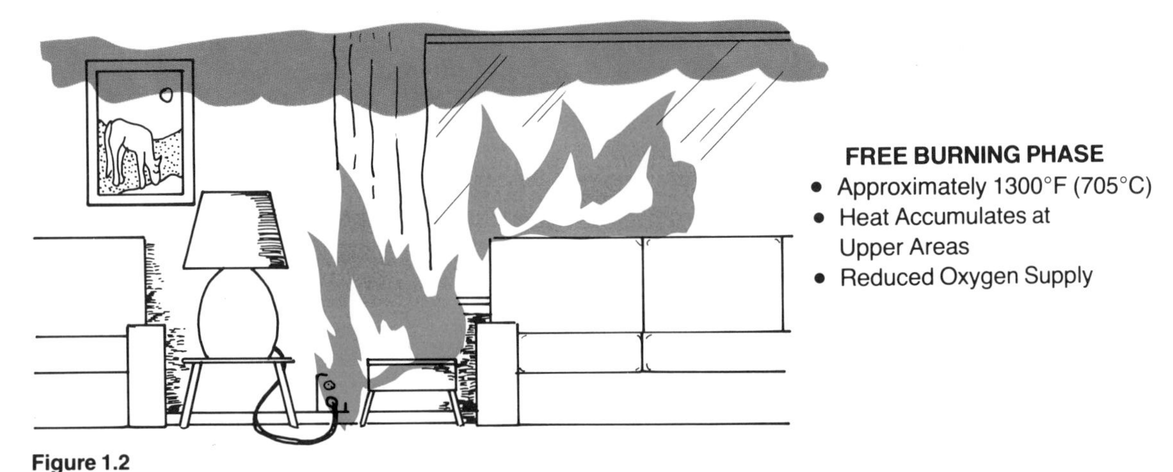

Figure 1.2

Smoldering Phase

In the third phase, flame may cease to exist if the area of confinement is sufficiently airtight. In this instance, burning is reduced to glowing embers. The room becomes completely filled with dense smoke and gases to the extent that it is forced from all cracks under pressure (Figure 1.3). The fire will continue to smolder, and the room will completely fill with dense smoke and gases of combustion at a temperature of well over 1,000°F (537°C). The intense heat will have vaporized the lighter fuel fractions such as hydrogen and methane from the combustible material in the room. These fuel gases will be added to those produced by the fire and will further increase the hazard to the firefighter and create the possibility of a backdraft.

Figure 1.3

PRODUCTS OF COMBUSTION

When a material (fuel) burns, it undergoes a chemical change. None of the elements making up the material are destroyed in the process, but all of the matter is transformed into another form or state. The products of combustion equal in weight and volume the weight and volume (although dispersed) of the fuel before it was burned. When a fuel burns there are four products of combustion: fire gases, flame, heat, and smoke.

Heat is a form of energy that is measured in degrees of temperature to signify its intensity. In this sense, heat is the product of combustion that is responsible for the spread of fire. In a physiological sense, it is the direct cause of burns and other forms of personal injury. Injuries caused by heat include dehydration, heat exhaustion, and injury to the respiratory tract, in addition to burns.

Flame is the visible, luminous body of a burning gas. When a burning gas is mixed with the proper amounts of oxygen, the flame becomes hotter and less luminous. This loss of luminosity is because of a more complete combustion of the carbon. For these reasons, flame is considered to be a product of combustion. Heat, smoke, and gas, however, can develop in certain types of smoldering fires without evidence of flame.

The smoke encountered at most fires consists of a mixture of oxygen, nitrogen, carbon dioxide, carbon monoxide, finely divided particles of soot and carbon, and a miscellaneous assortment of products that have been released from the material involved.

THE NATURE OF ACCUMULATED SMOKE AND GASES

The term *fire gases* refers to the vaporized products of combustion. The more common combustible materials contain carbon, which, when burned, forms carbon dioxide (CO_2) and carbon monoxide (CO). The principal factors that determine the fire gases that are formed by burning are the chemical composition of the fuel, the amount of oxygen present for combustion, and the temperature of the fire.

Fire gases and smoke dangerous

Firefighters may expect a closed building that contains a fire to be filled with smoke and hot gases. These gases and smoke are dangerous to the physical welfare of those entering the building, in addition to making it difficult to move through the building to locate the fire. In fact, a smoldering fire emits little or no flame to indicate its location, and even this is rendered obscure by a few feet of dense smoke.

As previously explained, the first gases that are normally produced by burning are principally water vapor and carbon dioxide. A lack of oxygen may later cause incomplete combustion and produce carbon monoxide. Carbon dioxide has no toxic effect upon a human body, but it acts as a suffocant in that it replaces the oxygen in the air. Carbon monoxide is an extremely toxic gas. If the atmosphere in a room contains as little as one-half of one percent of carbon monoxide, it will produce unconsciousness, if breathed, in a very short time. The presence of carbon monoxide is a situation against which firefighters must protect themselves by wearing protective breathing equipment.

It would be impossible to list all fire gases that exhibit toxic effects. Also, it would be impractical, since the amount present and the dose inhaled, ingested, or absorbed varies so widely. So, no attempt has been made to illustrate the exact dose necessary to cause harmful effects. Rather, the gases listed are the ones most commonly produced.

Senses often cannot identify toxic gases

Also, the characteristic odors, colors, symptoms, and signs mentioned may be partially or totally masked by the presence of many toxic and/or nontoxic products of combustion, thereby eliminating any possible sensory perception.

Most fire gases are toxic and will cause death or serious injury if inhaled in sufficient doses. The mechanism by which the toxic fire gases inflict their harmful effects varies somewhat between the various fire gases. In general, fire gases like carbon monoxide, carbon dioxide, and hydrogen cyanide will cause asphyxiation by disrupting the body's natural respiratory processes. Other fire gases such as hydrogen chloride (HCl), hydrogen sulfide (H_2S), sulfur dioxide (SO_2), ammonia (NH_3), and oxides of nitrogen cause excessive irritation in the respiratory tract, resulting in pulmonary edema. These irritations, some of which may appear minor, may in a few days develop into exceedingly disastrous complications and even death.

Fire Gas Characteristics and Effects

The following expresses the harmful effects one should expect if one were unable to or elected not to escape or evacuate a fire scene before being exposed to a harmful dose, or if one neglects to wear a self-contained breathing apparatus. Additionally, these gases are produced in most fires well in excess of the amount necessary to be considered deadly.

Toxic gases sink after fire extinguished

Many gases found in fire buildings have a vapor density greater than 1.0, meaning they are heavier than air at normal temperatures and would normally be found near the floor. However, because gases expand and become lighter as they are heated, the gases tend to rise. Most of the gases encountered in a fire building are hot since they are products of combustion; thus the greatest concentration of these gases are likely to be located in the upper portions of a fire room. If the fire is extinguished without complete ventilation, the heavy, heated gases will cool and return to the lower levels during overhaul. This phenomenon proves the necessity for continued use of protective breathing apparatus.

Ammonia is a colorless gas with a sharp, characteristic odor. It is a strong irritant of the eyes, nose, throat, and lungs. Symptoms and signs from exposure to this gas include a swelling of the eyelids, irritation of the nose and throat, coughing, vomiting, and difficulty breathing. It also causes irritation to moist skin.

Carbon dioxide is an odorless and colorless gas. When carbon dioxide is inhaled, it is absorbed by the bloodstream. As the concentration goes up in the bloodstream from inhalation, the blood's ability to remove carbon dioxide produced at the cells in the body tissue is diminished. *This procedure stimulates the individual's breathing rate, which can increase the inhalation of other toxic fire gases.* Moreover, this procedure reduces the amount of oxygen reaching the body tissue through the bloodstream. At slightly higher-than-normal levels of carbon dioxide in the bloodstream, the individual will experience headaches, dizziness, sweating, and mental excitement, in addition to increased lung ventilation. Still higher concentrations of carbon dioxide in the bloodstream can paralyze the respiratory center of the brain (caused by insufficient oxygen supply) and lead to death by asphyxiation.

Carbon monoxide is a colorless, odorless gas that acts as a chemical asphyxiant. Carbon monoxide combines with the hemoglobin in the blood 210 times faster than oxygen, thereby reducing the blood's ability to carry sufficient oxygen for supporting life functions of the body tissue. Carbon monoxide also hinders the removal of carbon dioxide from the bloodstream by the lungs. As the concentration of carbon monoxide in the bloodstream increases, the symptoms experienced will include shortness of breath, dizziness, mental confusion, impaired vision and hearing, and eventual collapse. The symptoms are not readily perceived by the afflicted person.

CO poisoning impairs judgment

Hydrogen chloride is a colorless gas. When it is inhaled, it reacts with the moisture of the body tissue in the respiratory tract to form hydrochloric acid, which causes a strong irritation. The primary harmful effect of an irritant gas such as hydrogen chloride is to cause edema at some level in the respiratory tract resulting in a blockage by fluids. Additionally, irritated tissue on the walls may swell up sufficiently to block the passage of air. Either of these conditions can lead to suffocation. Hydrogen chloride can also react with the water vapor in the atmosphere to form hydrochloric acid which has a detectable, pungent odor. The hydrochloric acid will condense in the atmosphere and form a white mist that will be visible. However, this white mist may be easily covered up or hidden by other products of combustion during a fire. Symptoms and signs include coughing, irritating pain in the respiratory tract, and breath holding.

Hydrogen cyanide is a colorless gas that has a faint odor of bitter almonds. It becomes chemically attached to the blood, where it prevents the transfer of oxygen from the blood to the cells of the body tissue. It causes death by asphyxiation and therefore is classified as a chemical asphyxiant. Signs and symptoms for exposure to hydrogen cyanide include headaches, dizziness, unsteadiness of gait, and a feeling of suffocation or nausea.

Hydrogen sulfide is a colorless gas but has a strong "rotten egg" odor. It has an irritating effect on the upper respiratory tract, leading to pulmonary edema. Pulmonary edema blocks the body's natural respiration process and leads to death by suffocation.

Oxygen deficiency is a condition of the fire atmosphere caused by the consumption of oxygen because of the combustion. Table 1.1 explains the symptoms for concentration of oxygen lower than the normal percentage in the atmosphere (21 percent).

TABLE 1.1
Oxygen Deficiency

Percentage of Oxygen in the Atmosphere	Symptom
17	Reduced muscular coordination, increased respiratory rate
15	Free burning ceases
12	Dizziness, headache, quick fatigue
9	Unconsciousness
6	**Death** within a few minutes from respiratory and concurrent heart failure

Nitrous fumes, produced by the burning of cellulose nitrates, are extremely dangerous. The term *nitrous fumes* covers a wide range of oxides of nitrogen. The oxides of nitrogen produce what may appear to be minor irritations, which may develop in a few days into exceedingly disastrous complications, even death.

There are two oxides of nitrogen that are dangerous: nitrogen dioxide and nitric oxide. *Nitrogen dioxide* is the most significant oxide of nitrogen, since nitric oxide would readily convert to nitrogen dioxide in the presence of oxygen and moisture. Nitrogen dioxide is a pulmonary irritant that has a reddish-brown color. When inhaled in sufficient concentration, it causes pulmonary edema. Nitrogen dioxide is an insidious gas in that its irritating effects in the nose and throat may be tolerated even though a lethal dose is being inhaled. Therefore, its hazardous effects from either its pulmonary irritation or chemical reaction may not become apparent for several hours after initial exposure.

Nitrogen dioxide a pulmonary irritant

Additionally, all oxides of nitrogen are somewhat soluble in water and react in the presence of oxygen to form nitric and nitrous acids. These acids are neutralized by the alkalis present in the body tissues and form nitrites and nitrates. These substances chemically attach to the blood and can induce nausea, abdominal pains, vomiting, and cyanosis (skin discoloration from oxygen deficiency in the blood), which can lead to collapse and coma. Also, nitrates and nitrites can cause arterial dilation, variation in blood pressure, headaches, and dizziness. The effects of nitrites and nitrates are secondary to the irritant effects of nitrogen dioxide but can become important under certain circumstances, causing delayed physiological reactions from exposure to the oxides of nitrogen.

Sulfur dioxide is a colorless gas that has a highly irritating property that makes it detectable below its lethal dosage. Sulfur dioxide is produced when materials containing sulfur are burned. This gas is suffocating but nonflammable. It is highly toxic in concentrations of 400 parts per million.

Other irritants: acetic acid, acetaldehyde, acetic anhydride, acrolein, formaldehyde, formic acid and furfural. These organic compounds are strong irritants with acid odors. Like most irritants, they are generally intolerable at levels below their toxic levels.

From the standpoint of flammability, carbon monoxide has certain characteristics that should be understood. The flammable range of carbon monoxide is between 12.5 and 74 percent, and the ignition temperature is approximately 1,200°F (650°C). This literally means that a mixture of 12.5 percent carbon monoxide with 87.5 percent air will flame when heated to its ignition temperature. In this condition, carbon monoxide is considered to be at its lower explosive limit. Also, a mixture of 74 percent carbon monoxide and 26 percent air will explode when heated to its ignition temperature. This condition is said to be at the upper

CO is flammable

explosive limit. This last percentage is seldom encountered in normal fire fighting. A smaller percentage of carbon monoxide mixed with other fuel gases and smoke does, however, set up a backdraft condition.

When firefighters enter a building, they may expect to find any or all of these products of combustion, depending upon what is burning. It is impossible to anticipate the nature of the smoke and fire gases until the material burning can be identified. If inspection surveys reveal what is likely to be burning, some degree of expectancy may be anticipated. Any exposure to any of the above toxic gases or fumes should be considered extremely dangerous. *Since it is extremely difficult to ascertain what gas is present, protective breathing equipment should always be worn in contaminated atmospheres.*

Wear SCBAs

A SMOKE-FILLED BUILDING

Some materials give off more smoke than others. Liquid fuel materials generally give off dense black smoke. Oils, tar, paint, varnish, molasses, sugar, rubber, sulfur, and many plastics also generally give off a dense smoke in large quantities.

The openings in a building may contribute to the accumulation of smoke in the areas not otherwise affected by the fire. If the fire occurs in one end of a room and if a vertical opening is in the opposite end, the smoke and gas often must travel the length of the room to make its escape. Thus, the entire room may be filled before the smoke shows on upper floors, especially if the fire occurs in a basement. If the wind is from the side where a fire starts, it may provide enough pressure inward to spread the smoke and gases throughout the building. Connections with adjoining buildings may afford the possibility of smoke appearing in those buildings rather than in the one in which the fire exists.

BACKDRAFT

Firefighters responding to a confined fire that is late in the free-burning phase or in the smoldering phase risk causing a backdraft or smoke explosion if the science of fire is not considered in opening the structure.

Proper ventilation avoids backdraft

In the smoldering phase of a fire, burning is incomplete because not enough oxygen is available to sustain the fire (Figure 1.4a). However, the heat from the free-burning phase remains, and the unburned carbon particles and other flammable products of combustion are just waiting to burst into rapid, almost instantaneous combustion when more oxygen is supplied. Proper ventilation releases smoke and the hot, unburned gases from the upper areas of the room or structure. Improper ventilation at this time supplies the dangerous missing link—oxygen. As soon as the needed range of oxygen rushes in, the stalled combustion resumes; and it can be devastating in its speed, truly qualifying as an explosion (Figure 1.4b).

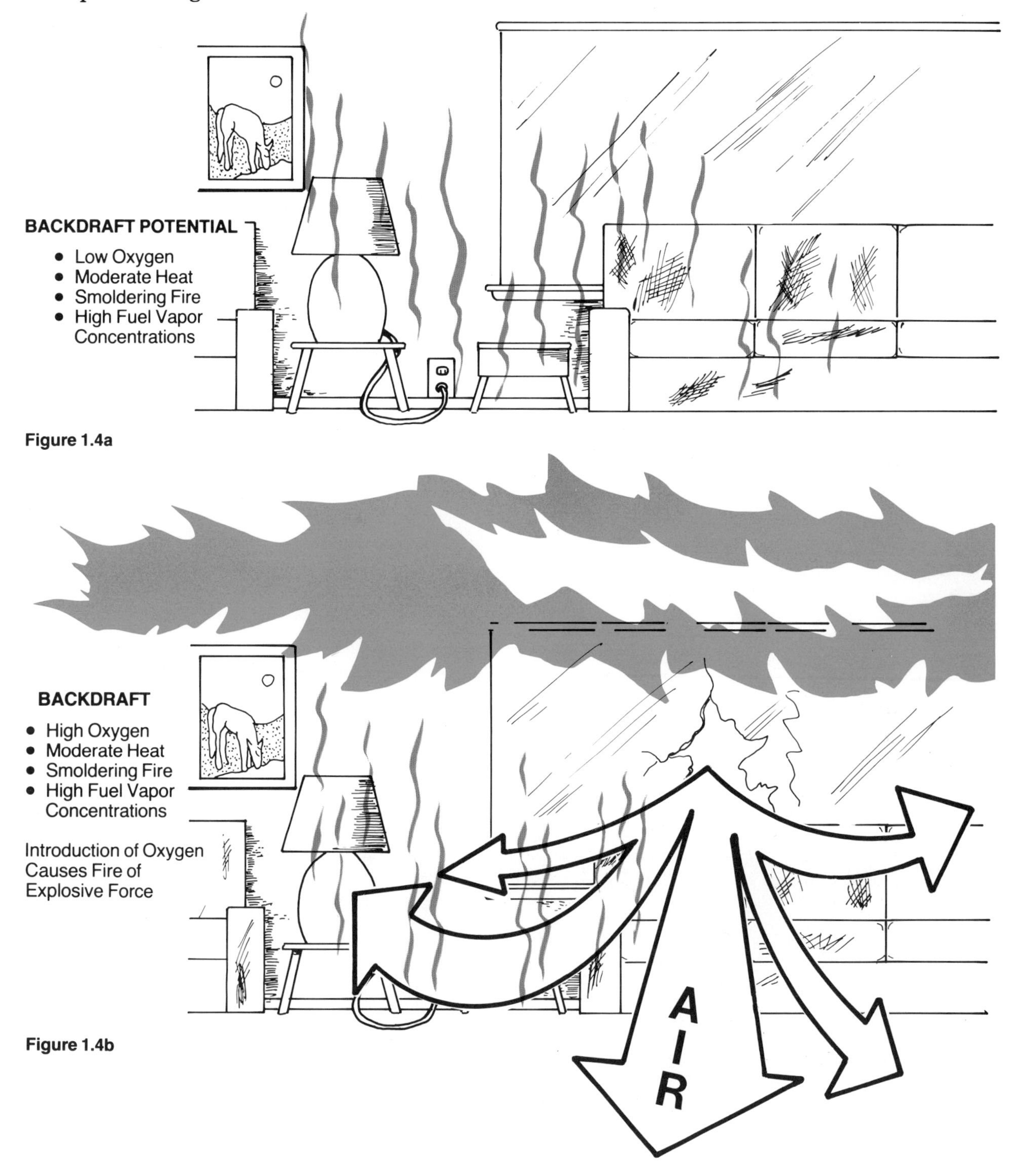

Figure 1.4a

Figure 1.4b

Combustion is related to oxidation, and oxidation is a chemical reaction in which oxygen combines with other elements. Carbon is a naturally abundant element present in wood, among other things. When wood burns, carbon combines with oxygen to form carbon dioxide (CO_2), or carbon monoxide (CO), depending on the availability of oxygen. When oxygen is no longer available, free carbon is released in the smoke. A warning sign of possible backdraft is dense, black (carbon-filled) smoke. When oxygen is reintroduced into the area, it combines with the carbon, and the smoke becomes less black, turning instead to yellow or grayish yellow. Then watch out! Figure 1.5 shows the result of opening a door at this stage.

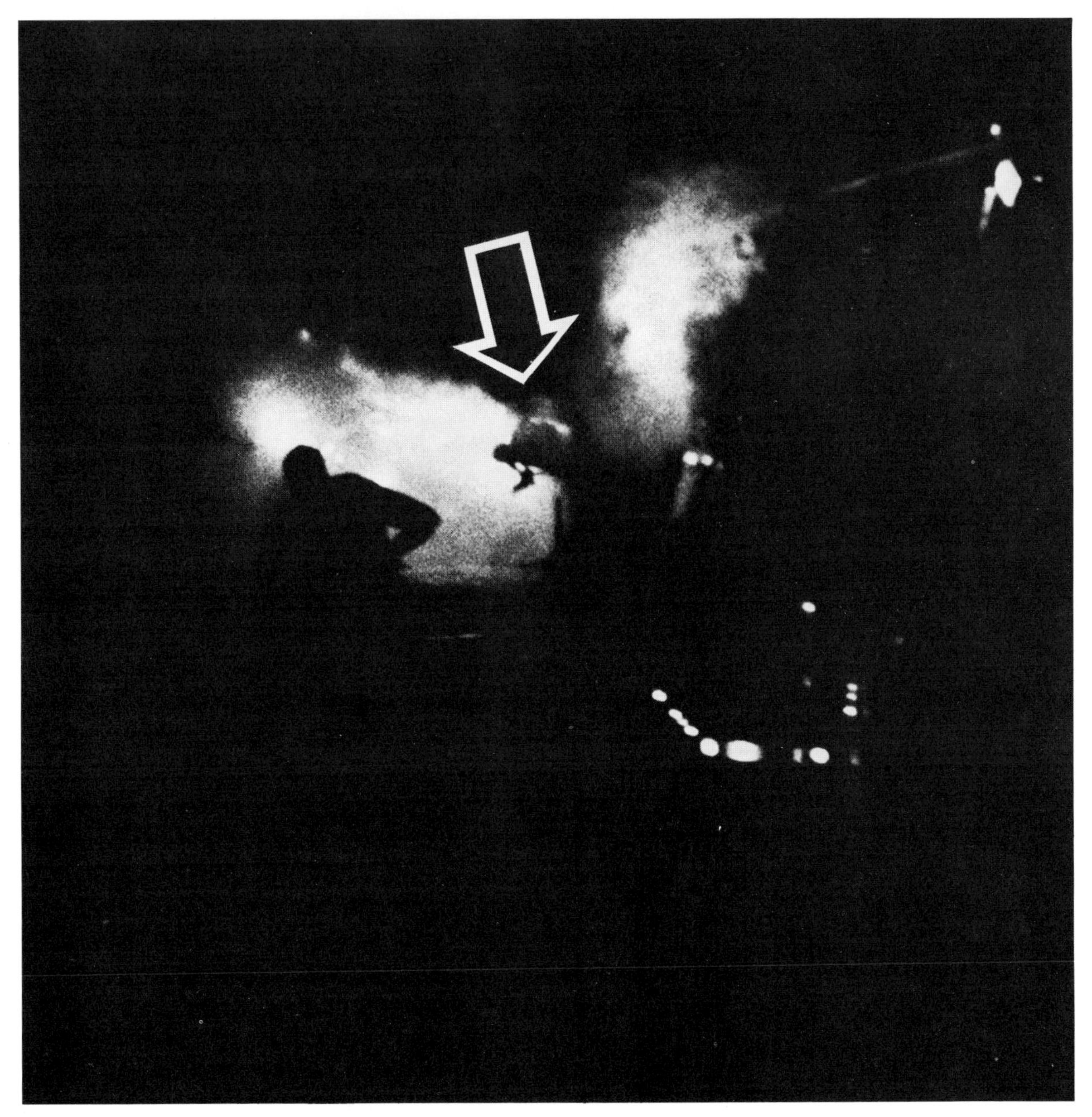

Figure 1.5 — This firefighter (arrow) was hurled through the air by the explosive force of a backdraft. *Nancy Engebretson*

Once again the importance of ventilation can be seen. The backdraft could be prevented by removing the unburned carbon particles and the heat, then the addition of oxygen would not complete the fire tetrahedron in a dangerous way. Instead, the cooling air would make subsequent fire fighting easier.

The following characteristics may indicate a backdraft or smoke explosion condition:

- Smoke under pressure.
- Black smoke becoming dense gray yellow.
- Confinement and excessive heat.
- Little or no visible flame.
- Smoke leaves the building in puffs or at intervals.
- Smoke-stained windows.
- Muffled sounds.
- Sudden rapid movement of air inward when opening is made.

This type of condition can be made less dangerous by proper ventilation. If the building is opened at the highest point involved, the heated gases and smoke will be expelled, reducing the possibility of an explosion.

TRANSMISSION OF HEAT

Heat can travel throughout a burning building by one or more of four methods, commonly referred to as conduction, radiation, convection, and flame contact. Since the existence of heat within a substance is caused by molecular action, the greater the molecular activity, the more intense the heat. A number of natural laws of physics are involved in the transmission of heat. One is called the *Law of Heat Flow;* it specifies that heat tends to flow from a hot substance to a cold substance. The colder of two bodies in contact will absorb heat until both objects are the same temperature.

Conduction

The process of conduction is generally associated with heat transfer, rather than with flame (Figure 1.6). Heat may be conducted from one body to another by direct contact of the two bodies or by an intervening heat-conducting medium. The amount of heat that will be transferred and its rate of travel by this method depends upon the conductivity of the material through which the heat is passing. Not all materials have the same heat conductivity. Aluminum, copper, and iron are good conductors. Fibrous materials, such as felt, cloth, and paper are poor conductors.

Liquids and gases are poor conductors of heat because of the movement of their molecules. Air is a relatively poor conductor. Certain solid materials when shredded into fibers and packed into batts make good insulation because the material itself is a poor conductor and there are air pockets within the batting. Double building walls that contain an air space provide additional insulation.

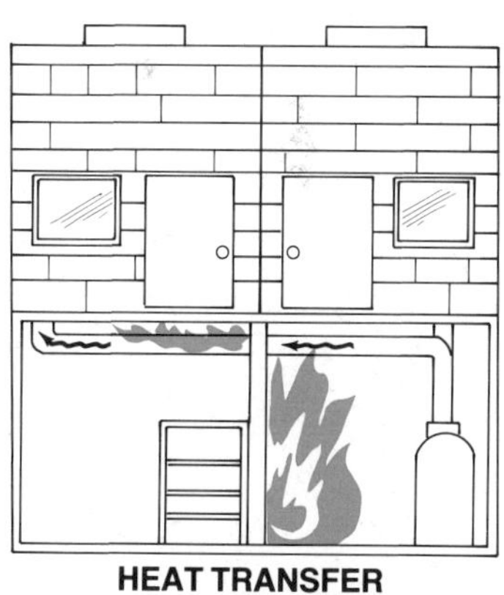

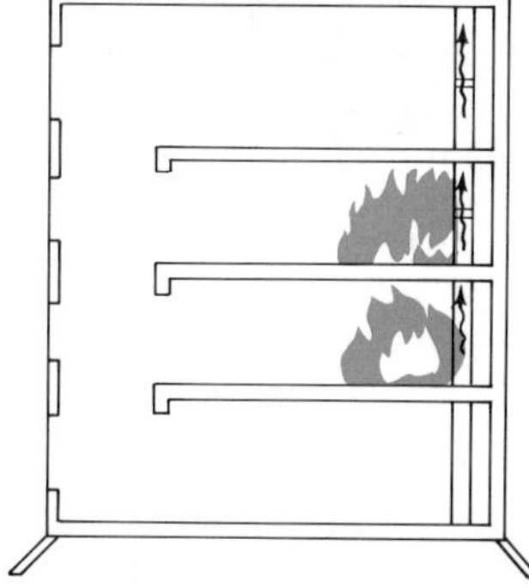

Figure 1.6 — Heat can be conducted through walls or floors by metal pipes igniting adjacent combustibles.

Radiation

The warmth of the sun is felt soon after it rises. When the sun sets, the earth begins to cool with similar rapidity. We carry an umbrella to shade our bodies from the heat of the sun. A spray of water between a firefighter and a fire will lessen the heat reaching the firefighter. Although air is a poor conductor, it is obvious that heat can travel where matter does not exist. This method of heat transmission is known as *radiation* of heat waves (Figure 1.7). Heat and light waves are similar in nature, but they differ in length per cycle. Heat waves are longer than light waves and they are sometimes called infrared rays. Radiated heat will travel through space until it reaches an opaque object. As the object is exposed to heat radiation, it will in return radiate heat from its surface. Radiated heat is one of the major sources of fire spread, and its importance demands immediate precautionary attention at points where radiation exposure is severe.

Figure 1.7 — Heat waves can radiate sufficient energy to cause ignition.

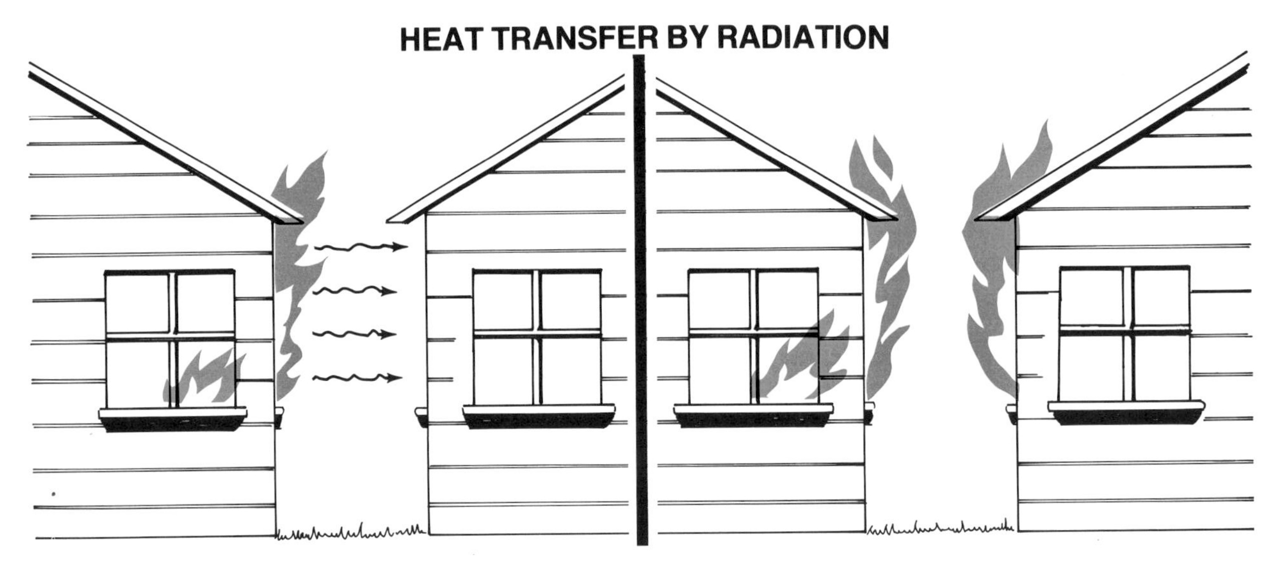

Convection

Convection is the transfer of heat by the movement of air or liquid (Figure 1.8). When water is heated in a glass container, the movement within the vessel can be observed through the glass. If some sawdust is added to the water, the movement is more apparent. As the water is heated, it expands and grows lighter; hence, the upward movement. In the same manner, air becomes heated near a steam radiator by conduction. It expands, becomes lighter and moves upward. As the heated air moves upward, cooler air takes its place at the lower levels. When liquids and gases are heated, they begin to move within themselves. This movement is different from the molecular motion discussed in conduction of heat and is known as heat transfer by convection.

Heated air in a building will expand and rise. For this reason, fire spread by convection is mostly in an upward direction, although air currents can carry heat in any direction. Convected currents are generally the cause of heat movement from floor to floor, from room to room, and from area to area. The spread of fire through corridors, up stairwells and elevator shafts, between walls, and through attics is mostly caused by the convection of heat currents and has more influence upon the positions for fire attack and ventilation than either radiation or conduction.

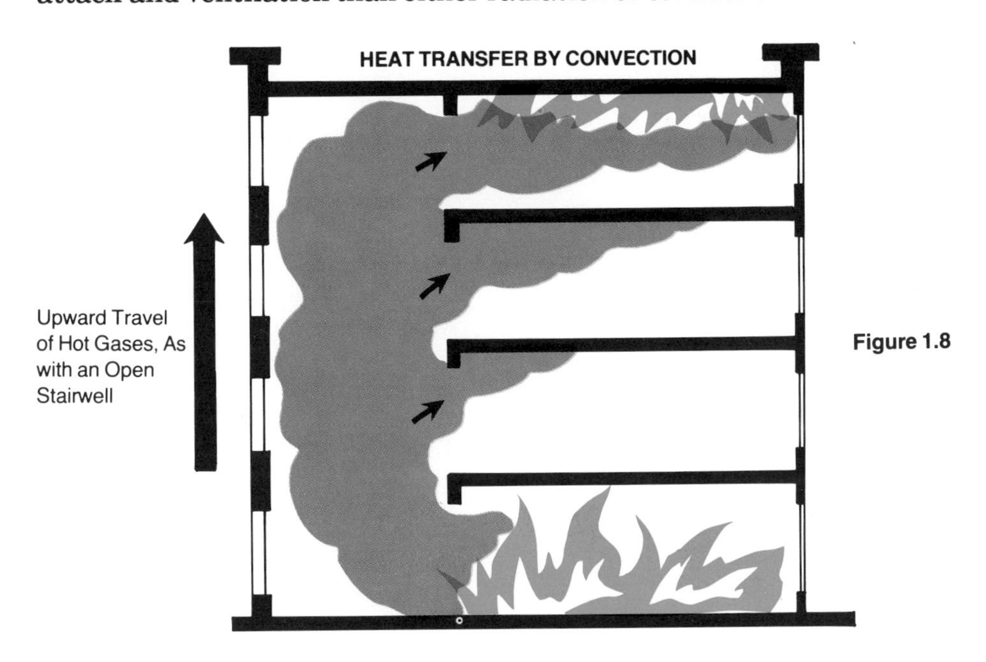

Figure 1.8

Direct Flame Contact

Fire also spreads along and through a material that will burn by direct flame contact (Figure 1.9). When a substance is heated to the point where flammable vapors are given off, these vapors may be ignited. Any other flammable material in contact with the burning vapors or flame may be heated to a temperature where it, too, will ignite and burn.

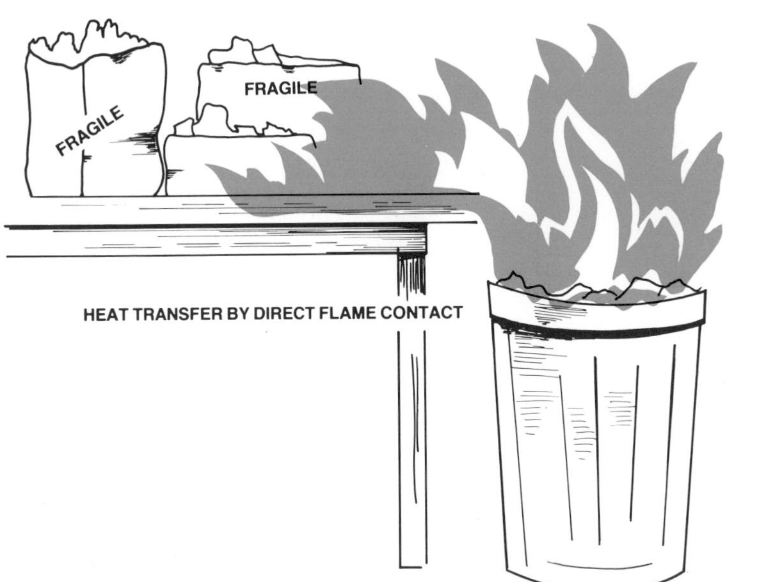

Figure 1.9

CONSTRUCTION FACTORS RELATING TO VENTILATION

2

NFPA STANDARD 1001

Fire Fighter II

4-10 Ventilation

4-10.2 The fire fighter shall identify the different types of roofs, demonstrate the techniques used to ventilate each type, and identify the necessary precautions.

This section of the manual is intended to encompass those aspects of construction with which a student of ventilation must be familiar and the nature of the many factors that will be considered in the ventilation decision. The degree of familiarity with construction features and the techniques of forcible entry that each firefighter possesses will have a direct bearing on the firefighter's effectiveness in the accomplishment of ventilation.

GENERAL BUILDING CONSTRUCTION FEATURES

Most modern buildings have continuous masonry foundations of concrete, brick, or stone upon which the building rests. The foundation walls that support frame construction may extend well above the ground. Exterior walls may be constructed of masonry, masonry veneer, metal, or wood frame (Figure 2.1).

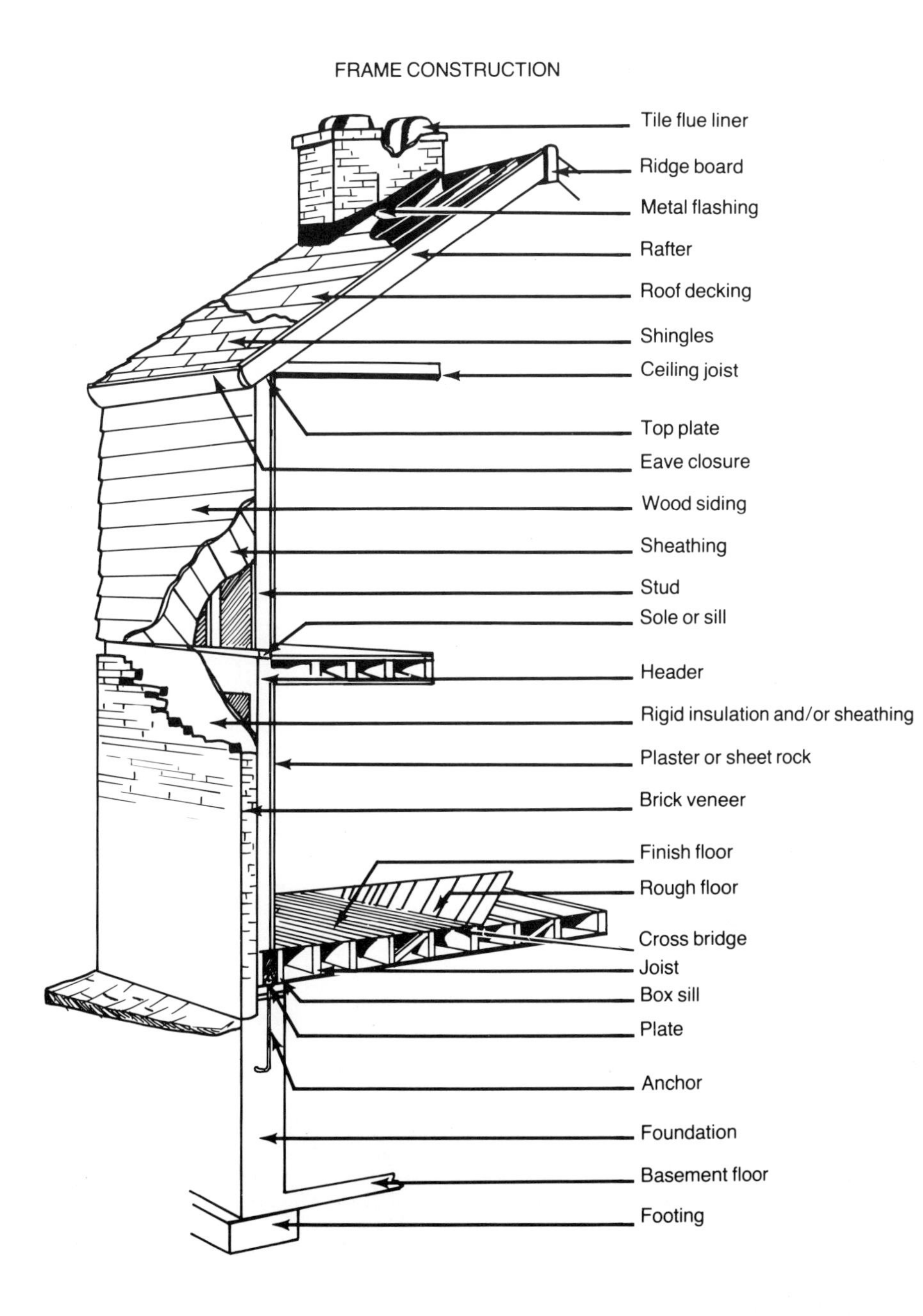

Figure 2.1 — The framing of a typical wooden-sided or masonry-veneered structure is the same for either.

Masonry exterior walls are the most desirable, from a fire protection point of view. They are usually eight to twelve inches thick depending upon the particular material used. Masonry veneer walls are essentially frame walls in which the supporting members are wood or metal with a veneer of one layer of brick or stone on the exterior to give the appearance of a solid brick or stone wall.

The upright wooden or metal supporting members in wood frame or masonry veneered walls are called *studs,* which are usually 2 x 4-inch size spaced at 16-, 18-, or 24-inch intervals. The presence of these studs creates hollow spaces in the walls through which fire can spread. Fire stops should be provided in these hollow spaces (Figure 2.2).

Frame walls are constructed of wood or wallboard sheathing fastened to studs. Over the sheathing is fastened the exterior siding, asbestos shingles, stucco, or other types of exterior finish. Frame walls also have hollow spaces in them that are created by the studs, which should also be blocked off with fire stops.

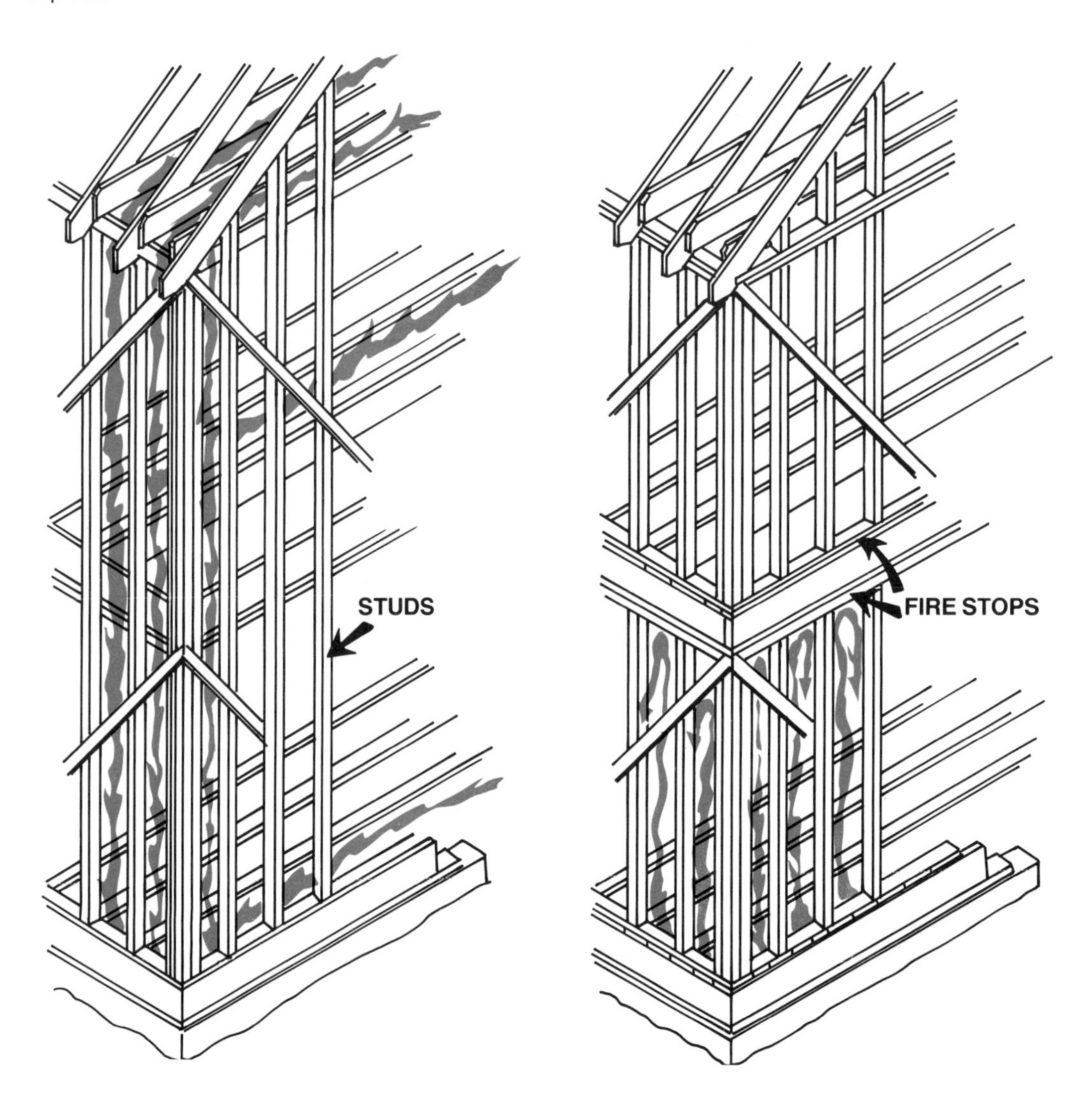

Figure 2.2 — Vertical fire spread through the walls of balloon frame structures (left) is easy and rapid. More modern construction (right) incorporates fire stopping to hinder such vertical fire spread.

Fire stops are construction members that are placed in walls, partitions, ceilings, between the studs at each floor level, and at the upper end of the stud channels in the attic. These fire stops cut off draft within the walls and help control the spread of fire and smoke. When fire stops are not used or the spaces are not filled with noncombustible insulation, it is possible for a fire originating in the lower levels to spread through the walls of the building before it is discovered. Modern construction uses top and bottom plates on studs to create a section of the wall before erecting. This fire-stopping method will prevent fire from traveling from floor to floor, but it can still travel the height of the partition or wall.

Fire walls are self-supporting walls designed to maintain structural integrity even in case of a major collapse of the structure (Figure 2.3). Fire walls should extend through and above combustible roofs to prevent the spread of fire in attics or on roofs. If fire walls are constructed properly, it should not be necessary to perform ventilation on the opposite side of the wall. However, if fire walls have unprotected openings, holes or poke-throughs, ventilation will have to be performed on the other side of the wall. Fire walls should be checked during routine fire department company building surveys and inspections in order to determine if the walls are constructed and maintained properly. The information obtained from company surveys and inspections should be used for pre-fire planning.

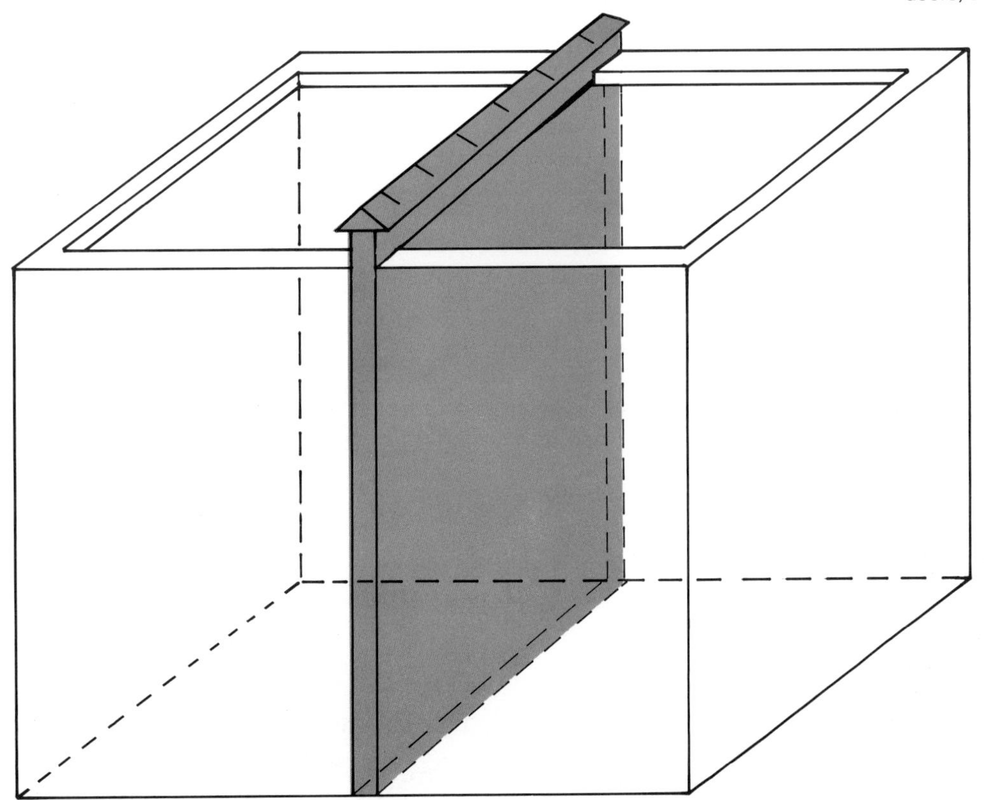

Figure 2.3 — A well-constructed fire wall prevents the spread of fire from one part of a building to another. The fire wall's integrity must not be broken by ductwork, doors, or similar openings.

PREFABRICATED STEEL BUILDINGS

Prefabricated steel buildings are constructed with a steel I-beam frame with thin sheets of steel or aluminum placed over the frame. Some of the older buildings have poles or columns in place of I-beams as a form of support. Insulation is placed inside the steel siding and dry wall is then placed over the insulation. It is important for the firefighter to remember that the roof is made in the same manner as the walls.

Many times, plastic skylights are placed on the roof. During fire situations, these plastic skylights have been known to melt out before the arrival of the fire department and thus aid the firefighter in ventilation.

Because the main component of these buildings is steel, they have been known to collapse when the internal temperature is 1,000°F (538°C) or more. At any time during a fire situation in which firefighters must be sent to the roof for ventilation, the use of aerial ladders or elevated platforms is recommended. This way, firefighters may work from these devices without the possibility of falling through a weakened roof. A firefighter who is sent onto the roof should first look for signs of discoloration and sagging of the roof and should always use a roof ladder, which is placed over the roof in order to distribute the firefighter's weight over a larger area. It is very important for the fireground officer to make a proper size up of the fire situation before sending firefighters to the roof for ventilation and inside for initial fire attack, in order to avoid possible injury to the firefighter.

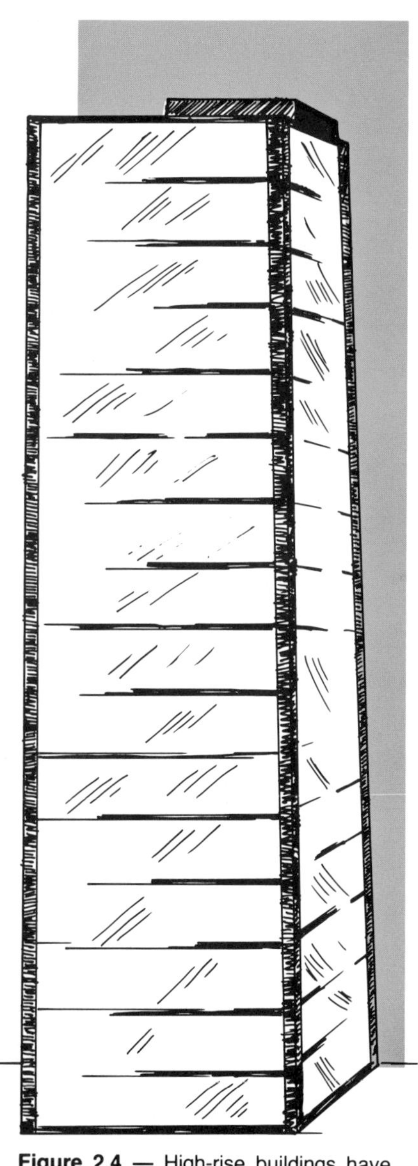

Figure 2.4 — High-rise buildings have special ventilation problems.

HIGH-RISE BUILDINGS

A major consideration in high-rise buildings is the danger to occupants from heat and smoke. Occupants of the building must be made aware of the reasons for, and the necessity of, proper fire prevention procedures and what actions they should take in the event of fire. They should be taught:

- The value of closed doors in preventing the spread of fire, smoke, heat, and gases.
- How to transmit a fire alarm.
- How to alert other occupants.
- The operation of fire extinguishers and standpipe equipment.
- How to prevent smoke from entering their rooms, by sealing cracks and closing vents.
- How to obtain fresh air if trapped in a room.

Architects have attempted to solve high-rise problems by fire-resistive construction, vertical enclosures, and compartmentation; but they are defeated by flammable, decorative materials and furnishings.

High-rise buildings are normally occupied by hospitals, hotels, apartments, or business offices. In any case, a great number of people may be exposed to danger if proper engineering, education, and enforcement is not provided.

Doors to stairway enclosures are frequently left, or propped, open, allowing smoke and heat to be drawn into hallways and rooms (Figure 2.5). Stairway enclosure doors must be automatic closing and must be kept closed.

In high-rise facilities, fires and smoke may spread rapidly through pipe shafts, stairways, elevator shafts, air handling systems, and other vertical openings. These shafts contribute to a "stack effect," creating an upward draft and interfering with evacuation and ventilation (Figure 2.6).

Elevators frequently become an undesirable means of transportation during fire fighting operations because of electrical failure, doors being held open on the fire floor because of activation of the electric eye mechanism by smoke, or the car being automatically recalled to the fire floor. Therefore, elevators must be equipped with a manual override system to provide safe operation by fire department personnel.

Figure 2.5 — Occupants sometimes create a potential smoke hazard by propping stairway doors open.

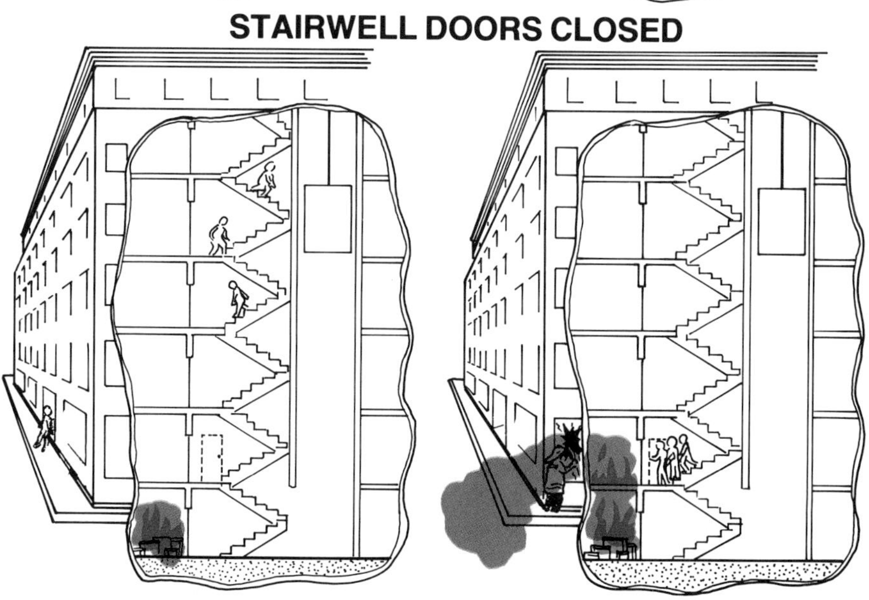

Figure 2.6 — Open shafts create upward drafts causing evacuation problems.

A haphazard approach to ventilation in a high-rise building fire, without specific plans for the effective use of manpower, equipment, and extinguishing agents, should never be attempted. Consideration must be given to the fact that the manpower demand for this type of building is approximately four to six times as great as required for a normal residence fire. In most instances, ventilation must be accomplished horizontally on the level involved by fire and immediately above the fire where smoke and heat have spread. Protective breathing equipment will be in great demand since, in most instances, every firefighter will have to be completely equipped for every eventuality. The problems of communication and coordination between the various attack and ventilating teams become more involved as the number of participants increases. Nozzlemen must be relieved as they reach their maximum heat-exposure limits.

STRATIFICATION OF SMOKE AND FIRE GASES

The creation of layers of smoke and fire gases on floors below the top floor of sealed multistory buildings is a relatively new phenomenon. Smoke and fire gases produced will accumulate at various levels until the building is ventilated. Pre-fire planning should include tactics and strategy that can cope with the ventilation and life hazard problems inherent in stratified smoke.

Heat escapes to the environment as smoke and fire gases travel through a building. The products of combustion rise through any vertical opening until their temperature is reduced to the temperature of the surrounding air. When this stabilization of temperature occurs, the smoke and fire gases form layers or clouds within the building.

Experience has shown that these dense smoke clouds form at a level below the top floor (Figure 2.7a). A classic example of this formation occurred in a 17-story sealed building. The fire was in the basement, and the dense smoke clouds formed on the tenth, eleventh and twelfth floors. The fire was extinguished before sufficient heat has built up to move the stratified smoke to the top floor. Ventilating this cooled smoke out of the top of the building was accomplished by creating controlled currents of air up the stairshafts and across the smoke-filled floors.

The mushrooming effect, which is usually expected on top floors, does not occur in tall buildings until sufficient heat is built up to move the stratified smoke and fire gas clouds that have gather on lower floors in an upward direction (Figure 2.7b).

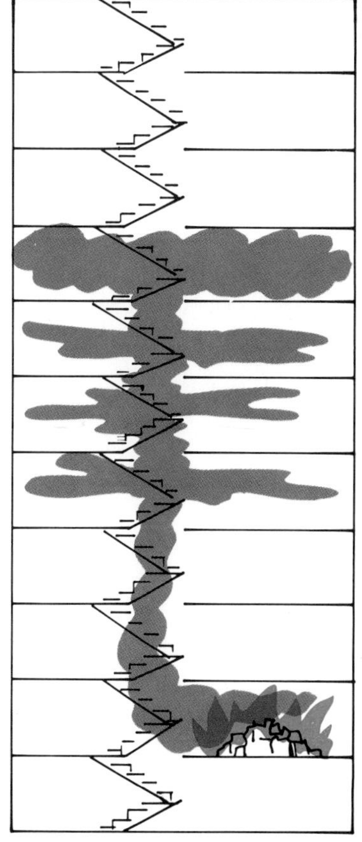

Figure 2.7a — In very tall buildings, mushrooming has a variant: The smoke and gases cool on their journey upward and layer themselves before reaching the topmost solid barrier. The smoke generally serves as a "lid" for other products of combustion. The level above the fire at which this can occur varies with a number of factors including fuel, intensity, and size of fire, types of smoke channels, height of the building, and weather.

MUSHROOMING

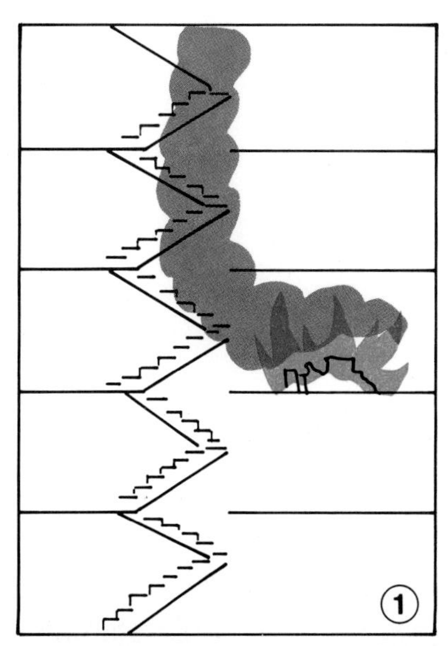

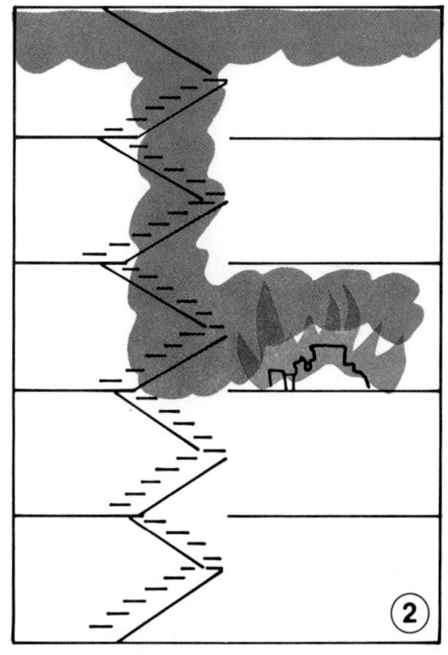

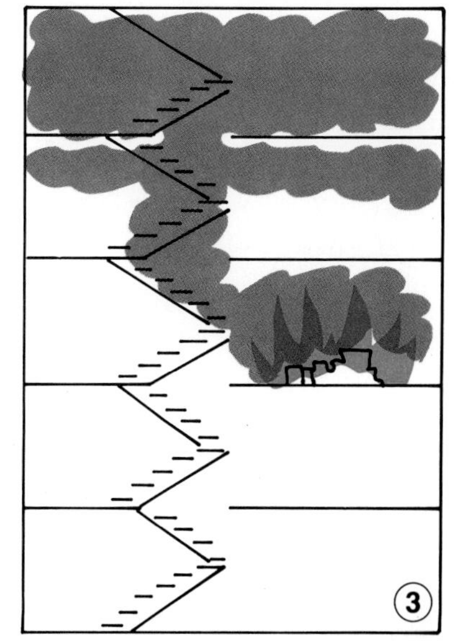

Figure 2.7b — When smoke and gases cannot escape vertically, they fill all available space from the top, down. Opening at the top will be necessary to prevent either backdraft or flashover if the mushrooming is advanced.

Preplan ventilation of high rises

Top ventilation of serious fires in modern tall buildings must be considered during pre-fire planning. In many buildings only one stairshaft pierces the roof. This vertical "chimney" must be used to ventilate smoke, heat, and fire gases from various floors. Before the doorways on the fire floors are opened and the stairshaft is ventilated, the uppermost door to this stairshaft must be blocked open in a very secure manner or removed from its hinges. Removal of the door at the top of the shaft insures that the door cannot close, which would allow the shaft to become filled with superheated gases, after ventilating tactics are started. Firefighters would then have no way of reopening the door except from the roof side. During ventilation, occupants and firefighters must not enter this shaft, because of the deadly chimney effect above the fire floor.

Timing and coordination of tactics are extremely important. Ventilation up the one stairshaft that pierces the roof must necessarily be delayed until all occupants above the fire floor are either evacuated or warned to stay out of the chimney and firefighters are known to be in a safe position. Once the decision has been made to ventilate this way, the stairway is useless for rescue operations, access, and escape.

Know stairway locations

Prior knowledge of stairway layout is imperative. Improper ventilation through a dead-end stairshaft can only seriously delay the extinguishment of the fire and increase the fire damage. Because the windows cannot be opened, common horizontal ventilation procedures may be impractical in newer buildings. Mechanical ventilation up a stairshaft, across a smoke-filled floor, and out through the roof by way of another stairshaft is the usual ventilation tactic in sealed buildings.

UNDERGROUND STRUCTURES AND WINDOWLESS BUILDINGS

Except in private dwellings, basements are required to have outside stairways, windows, or hoistways to provide firefighters access, except where the basement is sprinklered. Most basements and subbasements have outside entrances such as stairways, elevators, chutes, or windows. At the ground level, however, they may be blocked or secured by iron gratings, steel shutters, wooden doors, or combinations of these. Vault-type doors may be found on fallout shelters and on fur and bank vaults. If an outside entrance is provided, it is frequently sealed weathertight and locked for protection against burglars. All of these features serve to impede attempts at natural ventilation. An important factor that should be considered when a basement is involved in fire is its relationship to the main structure. Building features such as stairways, elevator shafts, pipe chases/shafts, laundry chutes, air handling systems, and other vertical openings contribute to the spread of fire and smoke to upper floors. Ventilation below street level is difficult in that it presents problems that are not encountered in ventilating above street level. Underground structures rarely provide an opportunity for normal ventilation. Basements usually require mechanical ventilation.

Figure 2.8 — Windowless structures are hard to ventilate.

Many buildings, especially in the business area, have windowless wall areas. Although windows are not usually a desirable escape route from buildings that are burning, they are most important during ventilation and rescue. Normal evacuation routes may frequently be blocked or made impassable during fires, and windowless buildings present a serious problem for trapped occupants. Windowless building designs also create an adverse effect on fire fighting and ventilation operations (Figure 2.8). It should not be overlooked, however, that windowless walls may retard the spread of fire. Some windowless structures have various forms of glass walls that cannot readily be opened, but they admit light to improve interior visibility. These walls may appear to be fragile, but they are as resistant to entry as are most masonry walls.

Good exit provisions and automatic sprinkler protection have been instrumental in eliminating much of the hazard to occupants and firefighters. The hazards are multiplied in windowless, fire-resistive buildings by the fact that ventilating the structure may be delayed for considerable time, allowing the fire to gain headway or creating backdraft conditions.

Problems inherent in ventilating this type of building are many and varied, depending upon the size, occupancy, configuration, and type of material from which the building is constructed. Windowless buildings usually require mechanical ventilation for the removal of smoke. From the standpoint of safety to occupants, it may be desirable to provide a means of removing smoke during the early stages of fire, yet the recirculation of smoke-laden air through the structure may increase the hazards, panic, and smoke damage. Air-handling systems should, therefore, be shut down as soon as possible. The means of doing this and the location of the controls should be noted in the preplan. Most buildings of this type are automatically air-conditioned and heated through ducts. This equipment, in combination with mechanical ventilation equipment, can effectively clear the area of smoke.

Familiarity with windowless buildings important

It is imperative that firefighters become thoroughly familiar with the design, contents, systems, and access routes in windowless buildings in their area. Management and occupants should be made aware of such items as will enhance the safety of occupants and assist in fire prevention and fire fighting efforts.*

ROOF CONSTRUCTION

Many designs and shapes of roof styles are used, and their names vary in each locality. Some of the more common styles are the flat, gable, gambrel, shed, hip, mansard, dome, lantern, and butterfly. Dwelling roofs are usually built of wooden supporting members called rafters, to which roof sheathing is fastened. To the roof sheathing is fastened or applied the roof-covering material. Roof coverings may be wooden shingles, composition shingles, composition roofing paper, tile, slate, or a built-up tar and gravel surface. The roof covering is the exposed part of the roof, and its primary purpose is to afford protection against the weather. The selection of a proper roof covering is important from a fire protection standpoint because it may be subjected to sparks and blazing brands if a nearby building should burn.**

Foamed-in-place roofs using synthetic foaming materials such as urethanes and isocynates are becoming more common. These roofs are being used for two reasons:

- Low maintenance cost
- High insulation qualities

Caution should be exercised when such roofs are encountered, because the flammability characteristics of some types are unknown. Also, the insulating qualities of the material could trap heat, leading to weakening or failure of the roof. When opening such roofs, a charged line should always be advanced to the roof in

---

*Standards and requirements for installed fire protection systems for underground structures and windowless buildings may be found in NFPA Standard No. 101, *Life Safety Code.*

**Classifications of roof coverings may be found in NFPA Standard No. 203M, *Roof Coverings.*

the event that the foam materials ignite and burn. Protective breathing equipment should *always* be worn for protection against toxic fumes released from burning foams.

Vertical (top) ventilation is directly related to roof construction and to vertical openings. A study of the more common types of roofs and the manner in which their construction affects ventilation practices is necessary to develop effective vertical ventilation procedures. The firefighter is concerned with three prevalent types of roof construction: flat, pitched, and arched. Buildings may be constructed with a combination of roof designs including these or other types, but an understanding of these three basic types is adequate for the purposes of ventilation as discussed in this manual.

Figure 2.9

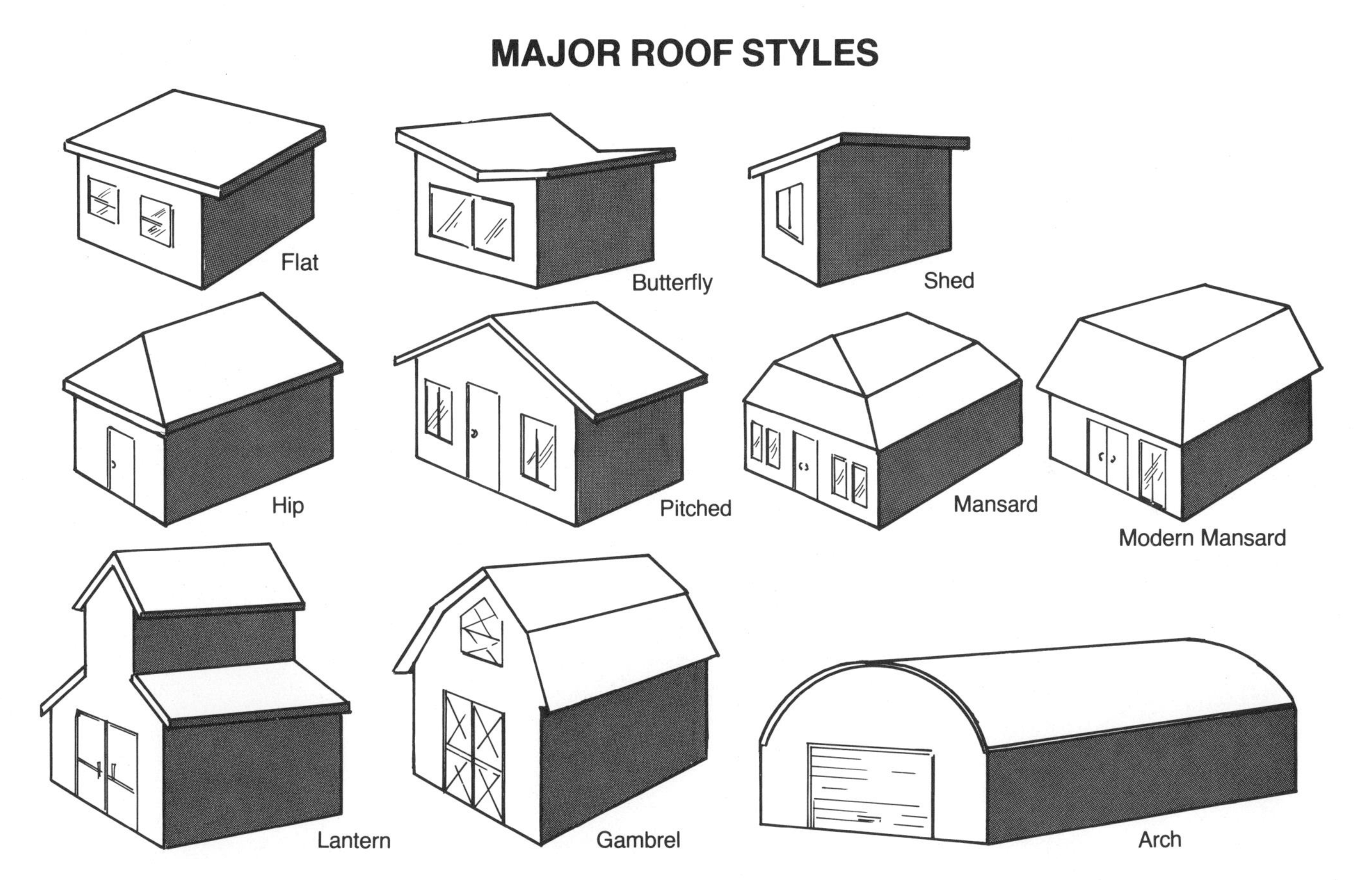

The Flat Roof

Flat roofs are more common to mercantile buildings, industrial buildings, multiple dwellings, and apartment complexes than to single-family dwellings. This type of roof ordinarily has a slight slope toward the rear of the building and is frequently pierced by chimneys, vent pipes, shafts, scuttles, and skylights. The roof may be surrounded and divided by parapets, and it may be expected to support water tanks, air-conditioning equipment, antennas, and other obstructions that may interfere with ventilation operations.

The structural part of a flat roof is generally similar to the construction of a floor, consisting of wooden, concrete, or metal rafters covered with sheathing. This sheathing is covered with a waterproof layer and an insulating material. Underneath the flat roof there is usually a concealed, possibly unvented, space that is formed when the ceiling is installed. These concealed spaces, or voids, complicate ventilation procedures and fire fighting to a great extent. Ventilators or pipe shafts may terminate in this space, and they can only be detected by complete inspections.

Many flat roofs are being constructed as shown in Figure 2.10. Such a roof can be dangerous during a fire because the chords of the trusses that support it are wood and burn through quickly. (The same type of truss is being used to support floors.)

Firefighters working on flat roofs should look for signs of heat or sagging of the roof. Serious consideration should be given to the use of roof ladders to distribute the firefighter's weight over a larger area to help prevent falling through a weakened roof.

Overhangs (Figure 2.11) are often added to flat-roofed buildings to give the apperance of a mansard roof. These overhangs form concealed spaces through which fire and smoke can spread quickly and unseen. Although they are not an integral part of the building, they might have to be opened. Overhanging cornices (Figure 2.12) might present the same problem.

Figure 2.10 — Parallel chord trusses such as these will burn through quickly and make the roof unsafe. Often there will be no indication that the roof will not support a person's weight.

Figure 2.11 — Overhangs can conceal and spread fire. They might need opening even though such opening probably will not ventilate the main fire.

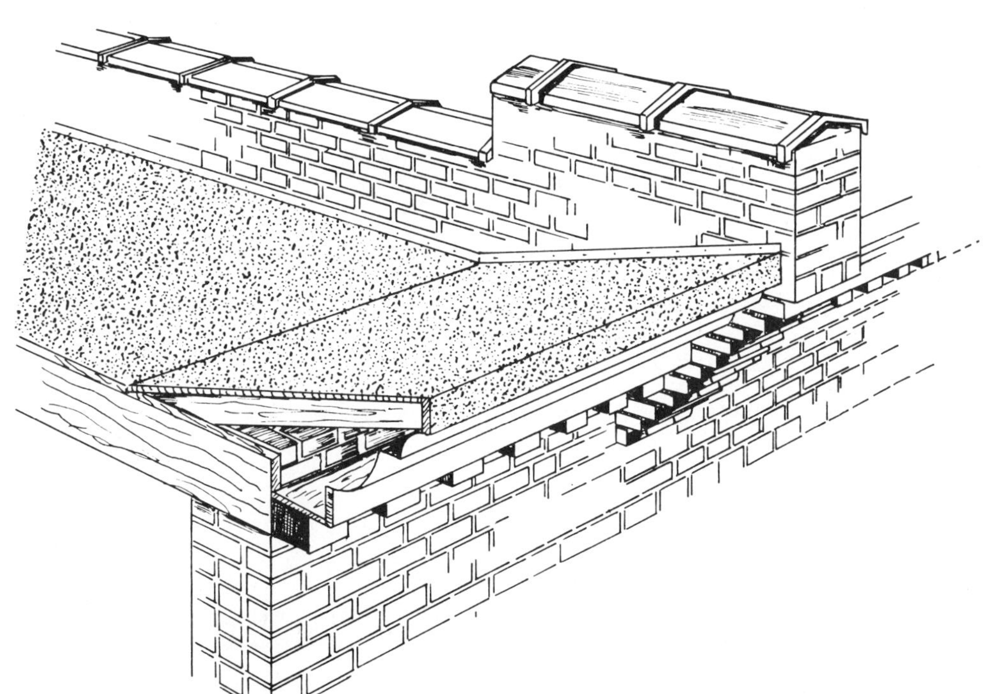

Figure 2.12 — Overhanging cornices might need opening if there is fire inside them.

The Pitched Roof

The pitched roof is any roof that is elevated in the center, thus forming a pitch to the edges. Pitched roof construction usually involves timber rafters or wooden or steel trusses that run from the ridge to a wall plate on top of the outer wall at eaves level (Figure 2.13). These rafters carry the sloping roof, which can be constructed of various materials. Ceiling joists are fastened to the wall plate and rafters so the assembly forms a series of triangles.

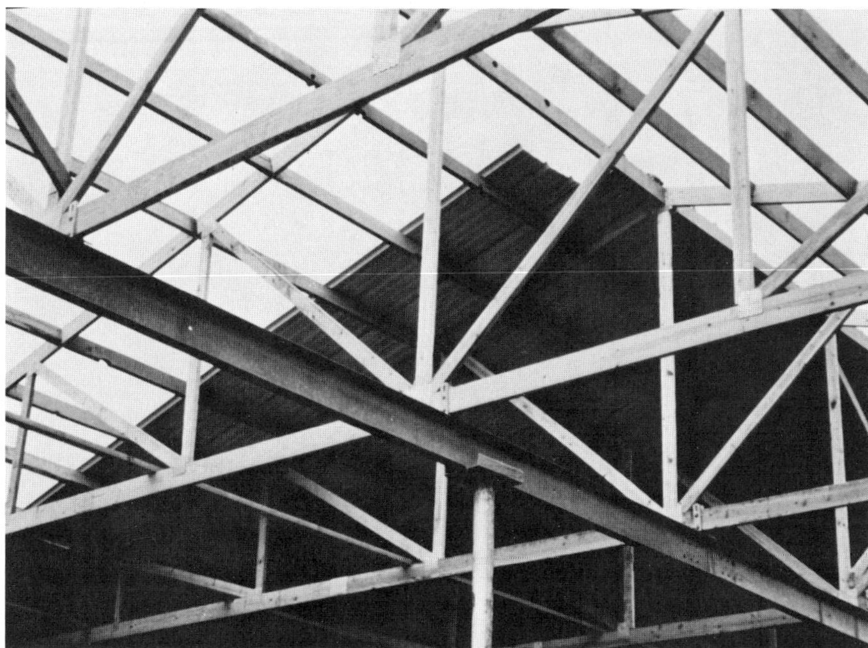

Figure 2.13 — The wooden pitched roof truss, like the wooden parallel chord truss, can burn through quickly, weakening the roof. Note that the only covering on this roof is a light metal sheathing.

There are many variations to this type of roof construction. In some cases, the ceiling may be omitted and the rafters made a part of a decorative interior plan. The usual trussed pitched roof is designed to cover a considerable span. Timber, metal, or, sometimes, concrete may be used as rafters. In the average home, the space between the roof members and the ceiling forms an attic space. These attic spaces are normally vented by louvers in each gable end, and entrance can be made to the attic by a small opening in the ceiling, usually in a hallway or closet. There may be a stairway leading into the attic. Insulation material that may or may not be combustible may be found between the framing under the roof or on top of the ceiling.

A pitched roof may be constructed to form a mansard roof. Instead of the roof running at a constant angle, there are two angles. One angle forms a steep pitch running from the eaves to a certain height, and the other produces a flatter pitch to the ridge or even to a flat roof.

The Arched Roof

Since their introduction, arched roofs have demonstrated many desirable qualities for certain types of buildings. The lower chord of the truss may be covered with a ceiling to form an enclosed cockloft or roof space. Such concealed, unvented spaces are definitely a hindrance to ventilation practices and contribute to the spread of fire.

Watch for concealed spaces

A trussless arched roof has been developed that is made up of relatively short timbers of uniform length. These timbers are beveled and bored at the ends where they are bolted together at an angle to form a network of structural timbers. This network forms an arch of mutually braced and stiffened timbers. Being an arch, rather than a truss, the roof exerts horizontal pressure in addition to vertical pressure. The result of these pressures is roof thrust, which is distributed to thrust supports. Trussless arch construction enables all parts of the roof to be visible to firefighters. A hole of considerable size may be cut or burned through the network sheathing and roofing at any place without causing collapse of the roof structure, since the loads are then distributed to less damaged timbers around the opening.

The Bowstring Truss Roof

In a bowstring truss roof, the arched member is dependent on all of the members holding it together. If one member fails, it is likely that the entire unit will fail. Many times truss units are connected to other truss units. When one unit collapses, it pulls down the roof and the other truss units, much like a series of dominoes, causing the front and rear walls to be pushed out. When the tension on a bowstring roof is released by the bottom cord burning through, pressure is applied to the outside walls and

Domino effect during collapse

can cause the collapse of the side walls as well as the roof. Bowstring truss roofs are commonly found in such occupancies as bowling alleys and supermarkets.

Figure 2.14a

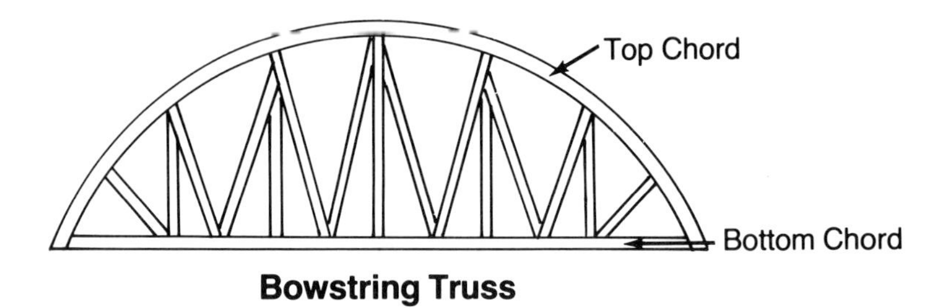

Figure 2.14b — The bowstring truss will fail from tension if either of its chords is broken.

ROOF OPENINGS

Firefighters should take advantage of natural roof openings wherever possible in accomplishing ventilation. In many instances, roof stairway access doors, hatches, skylights, ventilator openings, and elevator shafts can be used in vertical ventilation, thus eliminating the need to cut through roofs. Procedures of opening the roof are covered in a following section of this manual.

VENTING DEVICES

The need for smoke vents came as a result of construction and lighting improvements that enable large buildings to be built with windowless exterior walls and without supporting interior walls. If these large structures are not properly vented, curtained, and otherwise protected by permanently installed detection and extinguishing systems the entire contents are vulnerable to the spread of fire and smoke contamination.

Vents are not intended to be a substitution for fire-detecting or extinguishing systems, but they have proven effective in releasing the heated gases, reducing smoke damage, and retarding the spread of fire. The following information on the various types of vents and other construction features is provided as a basis for identification and familiarity.

Curtain boards are used to trap and retard the spread of smoke and heat (Figure 2.15). In addition to performing these two functions, curtain boards, used in conjunction with a vent, serve to increase the height of the stack. In the theory of draft, the effectiveness of the vent will be increased proportionately to the difference between the temperature of the air in the stack and the outside air. The higher stack created by the curtain board confines the heated air, keeping it from spreading out, thus increasing the temperature in the stack.

Curtain boards should be constructed of noncombustible material and should extend from the roof or ceiling to a height above the floor necessary to restrict the spread of heat and smoke horizontally through the open areas of the building. The areas covered or protected will generally be governed by local codes.

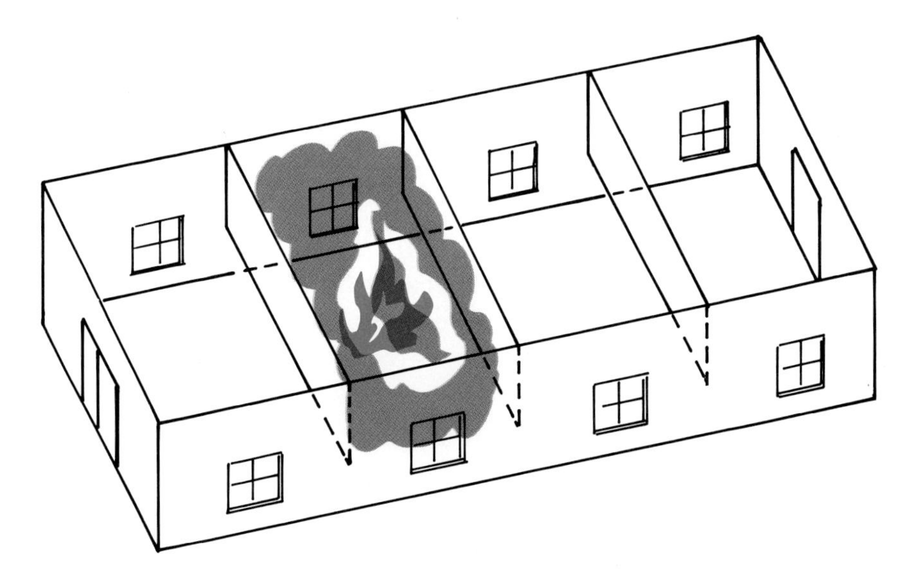

Figure 2.15 — Curtain boards are extended from the ceiling of a large open space to confine heat and smoke.

Monitors

Monitor vents are usually rectangular projections through the roof of single-story buildings. The monitor may have metal, glass, wired glass, or louvered sides. The type with glass sides depends upon the glass breaking to provide venting in case of a fire. The type with solid walls should have at least two opposite sides hinged at the bottom and held closed at the top with a fusible link arrangement to facilitate their opening in case of fire.

Continuous Gravity Vents

Continuous gravity vents are the unbroken narrow slot openings similar to those frequently found along the gables of pitched roofs. They are usually covered with weather hoods. If they are equipped with temperature control shutters, they should be of the type that automatically opens in case of fire.

Unit Type Vents

Unit type vents are normally constructed of metal frames and walls and operate by a hinged damper controlled either manually or automatically. They may be found in various sizes from 4 by 4 feet to 10 by 10 feet, distributed throughout the roof. Their location and number are determined by the requirements of the occupancy.

Automatic Heat and Smoke Vents

Note automatic vents during preplanning

Firefighters should become acquainted with methods of operation and the manual release mechanisms of automatic heat and smoke-venting devices. When preplanning structures, the location of the vents and methods of manual release should be noted on the preplan. Attempts to use forcible entry practices on automatic venting devices can be extremely dangerous because of spring loading or other operating systems. Additionally, attempts to force such vents can do extensive damage to their operating mechanisms as well as create a very dangerous situation for the firefighter. When smoke and heat vents are opened, mechanical smoke ejectors may improve the venting process.

Skylights that contain ordinary glass are effective vents since the temperature of a fire can cause the glass to break and fall out. Skylights equipped with wired glass have little or no value as automatic vents unless the frame is counterweighted and secured with a fusible link arrangement that will automatically open them in the event of a fire, or unless the entire unit is removed by firefighters.

Air Conditioning Systems

Air conditioning systems can be adapted for limited use in the control of smoke. These systems are not made to handle high-temperature smoke and fire gases, but they can prove to be an auxiliary method in smoke control. If the air conditioning system is able to create a greater pressure in the areas uninvolved in fire, this pressure will be exerted toward the fire areas, reducing smoke migration into the uninvolved areas. (See NFPA Standard No. 90A, *Air Conditioning and Ventilating Systems.)*

AUTOMATIC SPRINKLER HEADS

Activated sprinkler heads discharging water after a fire has been controlled tends to retard the ventilation of smoke. *However, automatic sprinklers must be left on until complete extinguishment is certain, because under smoky conditions hidden fires could still be present.*

Automatic smoke and heat vents may not automatically operate when activated sprinkler heads are discharging close to them. The sprinklers may control the fire and heat development to a level that would not activate the automatic smoke and heat vents by melting the fusible link.

EXTERIOR WALL WINDOWS

Exterior wall windows that reach near the ceiling may be of great value as vents. Windows that are lower are not as effective, since the hottest air will flow to the ceiling. The top windows on the leeward side of the building should be opened first, if conditions will permit horizontal ventilation from the point of their location.

OCCUPANCY AND CONTENTS

Occupancy and contents important considerations

The type of occupancy of a building determines the degree and nature of the hazard, along with the fire potential that may exist. Occupancy hazards are equally as important as construction features in relation to the study of ventilation. Special factors that will have a direct bearing on the ventilation procedure include:

- Number and location of personnel who can be expected to occupy the building.
- The extent to which escape routes may be blocked or rendered inaccessible by the fire or smoke, and alternative routes available.
- The quantity and type of combustibles that make up the contents.
- The location of these combustibles.
- The toxicity and explosion hazards of the contents.
- Fire-detecting and extinguishing systems available.

VENTILATION CONSIDERATIONS AND DECISIONS

A fire officer faces a many-sided problem when evaluating a ventilation situation. All of the officer's knowledge, training, and experience must be used to arrive at the best possible solution. Fires are not predictable, in that the variables cannot be completely controlled. Decisions must be based on those indicators with which the officer is familiar.

REQUIREMENTS FOR VENTILATION

Decisions to be made before ventilating

After all indicators have been closely observed and evaluated, the requirements for a plan of attack must be considered. Before a fire officer directs or orders a ventilation operation, a series of decisions must be made that will, by the nature of fire situations, fall into the following order:

First Decision

Is there a need for ventilation at this time?

The answer to this question must be based upon the life hazard and the extent of the fire, heat, smoke, and gas conditions within the structure.

Second Decision

Where is ventilation needed?

The answer to this question involves:

- Exposures
- Extent of the fire
- Location of the fire
- Wind direction
- Construction features of the building
- Vertical openings
- Horizontal openings

Third Decision

What type of ventilation should be used?

The answer to this question may be derived from a fire officer's knowledge of the following three methods of ventilation:

- Providing an opening for the passage of air between interior and exterior atmospheres.
- Using the application of water fog and the expansion of water into steam to displace contaminated atmospheres.
- Using mechanical (forced) ventilation.

Recent research has indicated that modern energy conservation policies may create ventilation problems since insulation blown into existing buildings may cause flashover to occur much faster. This is because the insulation will retain heat much better and will thus raise the temperature of combustibles in the fire area to ignition temperature much more quickly than is normally anticipated. Therefore, the need for ventilation is increased and must be accomplished much sooner than has been practiced in the past. Insulation installed over roof coverings of fire-rated roof construction will effectively retain heat and may reduce the fire rating drastically, causing premature roof failure.

Pre-fire inspection should note the roof construction and where extra insulation has been added to existing roofs and attic areas, so personnel performing ventilation may be aware of possible problems.

Firefighter safety paramount

When a fire officer determines the need for ventilation, the precautions that may be necessary for the control of the fire and the safety of the firefighters must also be considered. There is a need for protective breathing equipment for respiratory protection, and charged hoselines should be provided during the ventilation process. The possibility of fire spreading throughout a building and the danger of exposure fires are always present.

VISIBLE SMOKE CONDITIONS

Smoke conditions will vary according to how the burning has progressed. A free-burning fire must be treated different from one in the smoldering stage. A fire that is relatively small in size is frequently mistaken for a large generalized fire because of the large volume of smoke. Smoke accompanies most ordinary forms of combustion, and it differs greatly with the nature of the substances being burned and the amount of available oxygen. In addition to the gaseous products of combustion, other constituents such as tar, unburned carbon, and ash are drawn upward by the draft created by the heat of the fire. The density and color of the smoke is in direct ratio to the amount of suspended particles. A fire that is just starting and is consuming wood, cloth, and other ordinary furnishings will ordinarily give off gray white or blue white smoke of no great density. As the burning progresses the density may increase, and the smoke may become darker because of the presence of large quantities of carbon particles.

Smoke color unreliable indicator

Black smoke is usually the result of burning hydrocarbon such as rubber, tar roofing, oil, or plastics. It has been said that brown smoke may indicate nitrous fumes and that gray yellow smoke is a danger signal of an approaching backdraft. A firefighter should remember that the chemicals that smoke may contain can only be determined by chemical analysis. *Although the color of the smoke may be of some value in determining what is burning, smoke color is not always a reliable indicator.*

The point at which the smoke is escaping may offer some indication of a possible solution to the ventilation problem. If smoke is coming out of a lower story, it may indicate that openings are blocked above the point where smoke appears (Figure 3.1). Several possibilities must be considered when determining where to ventilate. If smoke is coming through only one end of the building, it may indicate the location of the fire or it may be the result of wind direction (Figure 3.2).

When a fire is confined and has gained much headway, the smoke may be coming out around the roof, through openings, around skylights, penthouses, scuttle holes, and even through small openings in the walls. It may be lazily drifting out into the atmosphere or it may be coming out with a great deal of force. Tests show that only a slight positive pressure exists inside a burning building. The speed with which smoke emerges is, however, an indication of inside conditions.

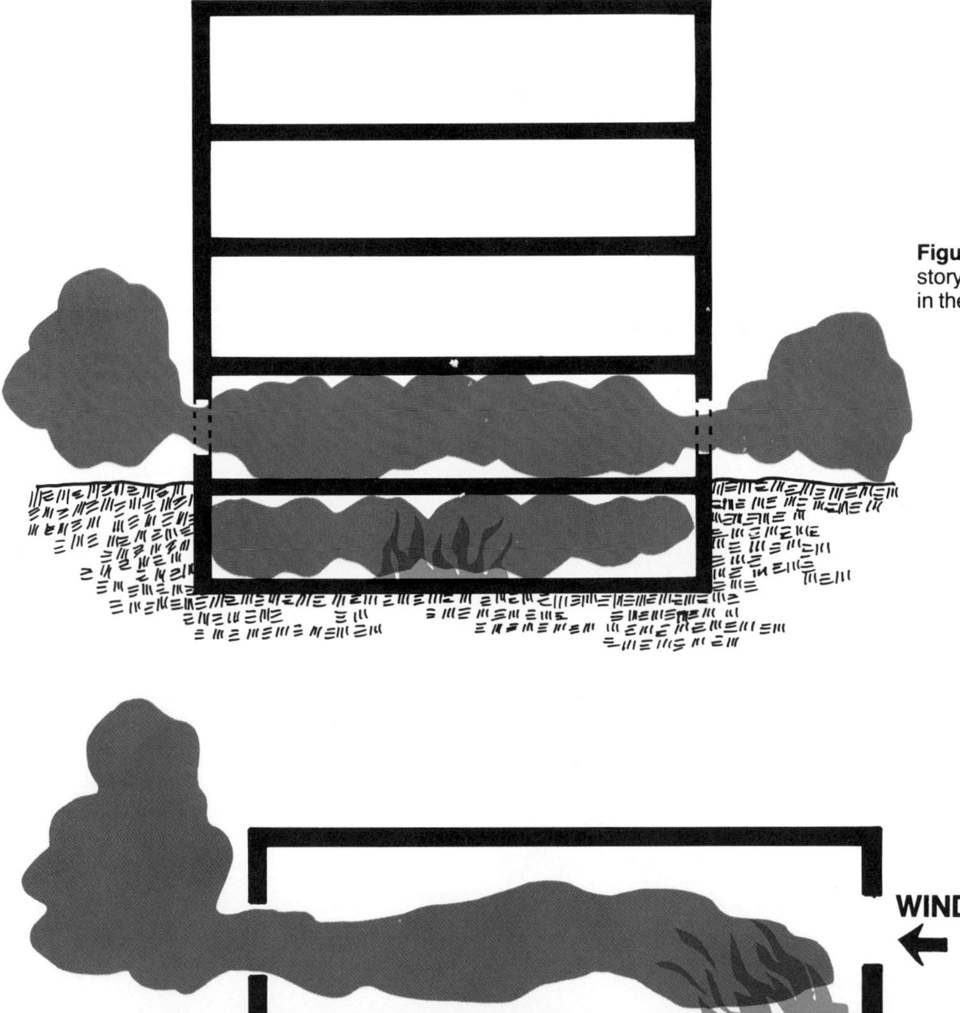

Figure 3.1 — Smoke coming from a lower story could indicate that openings higher in the building are blocked.

Figure 3.2 — The place where smoke appears does not necessarily indicate the location of the fire. Wind currents, building layout, and closed doors can affect smoke travel.

SMOKE ODORS

It is sometimes quite easy for firefighters to classify what is burning by distinctive odors, especially during the early stages of the fire. Smoke from burning rubber, food, electrical insulation, rags, pine wood, feathers, or grass all have a characteristic odor and will permit a firefighter to determine what kind of material is burning.

PROVIDING FIRE CONTROL

Before ventilating a building, a fire officer must provide manpower and adequate equipment for fire control. The fire may immediately increase in intensity when the building is opened. These resources should be provided for both the building involved and other exposed buildings. As soon as the building has been opened to permit hot gases and smoke to escape, the next requirement is to reach the seat of the fire for extinguishment. Entrance should be made into the building through a door, window, or some such opening as near the fire as possible if wind direction will permit. It is at this opening that charged hoselines should be in readiness in case of violent burning or an explosion. Charged hoselines should also be made available at other points where openings are made to protect buildings that are likely to be endangered because of their exposure to the one involved (Figure 3.3).

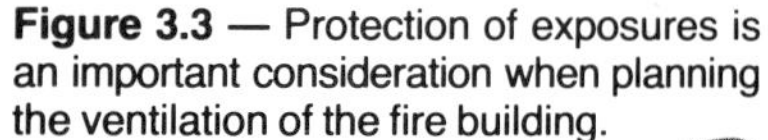

Figure 3.3 — Protection of exposures is an important consideration when planning the ventilation of the fire building.

Hose streams should not be directed into ventilation openings created by the fire or firefighters because this will defeat the ventilation process and may endanger persons inside the building. In addition, this action may force firefighters in the building to abandon their attack positions and allow the fire to extend into uninvolved portions of the building.

HEAT CONDITIONS AND FIRE SEVERITY

Smoke conditions cannot always be well observed, especially at night. As a fire progresses, high temperatures develop. These high temperatures will still be present even in the third phase, or smoldering state.

Firefighters may obtain an indication of the intensity of the heat by feeling walls, doors, or windows. A general heated condition will indicate the interior temperature. Hot spots will indicate the probable travel of the heat or the location of the fire. A lower-level fire may sometimes be located by a hot spot on the floor above it. Where an involved building is one of a connected series in a business block, an investigation of the interior walls or exterior windows of adjoining buildings may disclose the location of a fire. *A little time for consideration is worthwhile when determining proper ventilation procedures. The point and method of attack should be well established.*

Time taken is time gained

LIFE HAZARDS TO OCCUPANTS

Dealing with the danger to human life is of utmost importance. Certain fire conditions may suggest that ventilation come first to draw away heat and smoke, or that the spreading flames must be attacked immediately; sometimes both must be done simultaneously. All variables cannot be discussed here, but the point is that the first consideration is the safety of occupants. The life hazard is generally reduced in an occupied building involved by fire if the occupants are awake. If, however, the occupants were asleep when the fire developed and are still in the building, either of two situations may be expected: first, they may have been overcome by smoke and gases; second, they might have become lost in the building and are probably panicky. In either case, proper ventilation will be needed in conjunction with rescue operations (Figure 3.4).

POTENTIAL HAZARDS TO FIRE SERVICE PERSONNEL

In addition to the hazards that endanger occupants, there are potential hazards to firefighters and rescue workers. The type of structure involved, whether natural openings are adequate, and the need to cut through roofs, walls, or floors (combined with other factors) add more problems to the decision process.

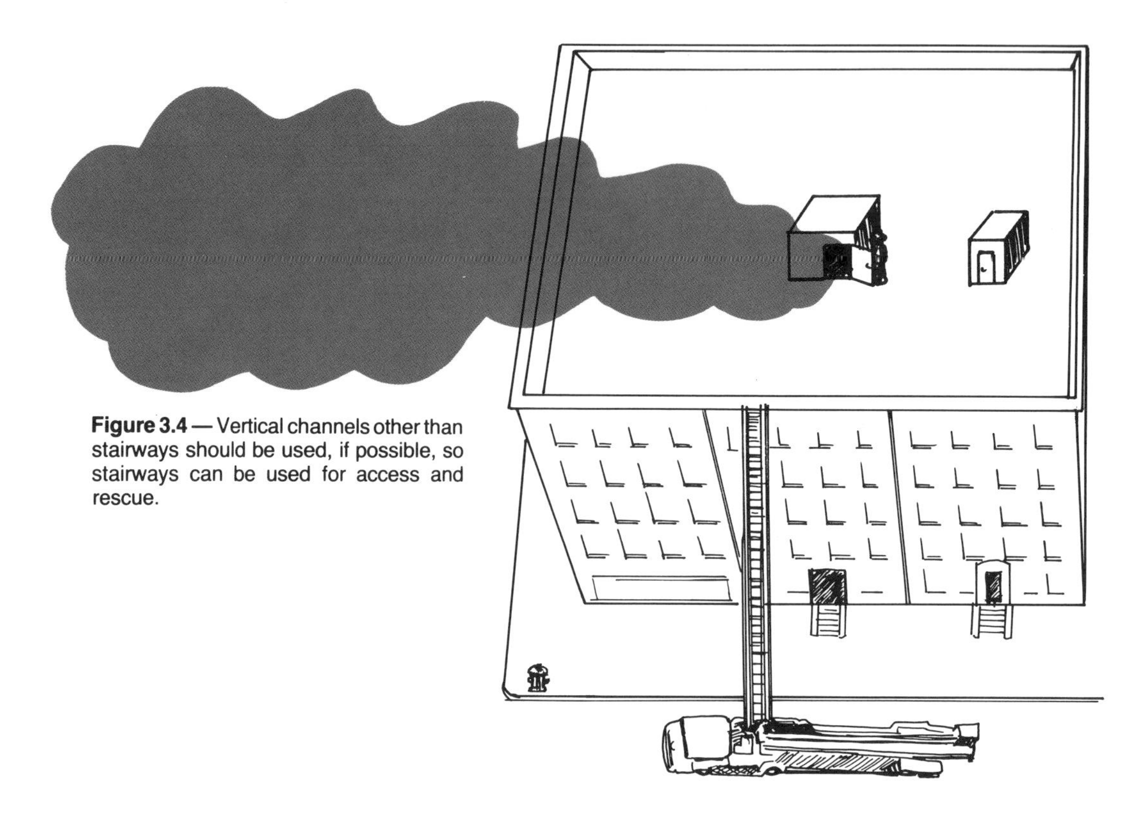

Figure 3.4 — Vertical channels other than stairways should be used, if possible, so stairways can be used for access and rescue.

The hazards to be expected in a closed building that contains a smoldering fire are life hazards, fire hazards, and backdraft hazards. A review of the nature of the smoke and gases during a fire reveals the following potential hazards:

- The obscurity caused by dense smoke.
- The presence of poisonous gases.
- The lack of oxygen for breathing.
- The presence of flammable gases.
- Serious damage to the structure.

Since a fire may raise the temperature in a room to well over 1,000°F (538°C), a firefighter who has a basic knowledge of the ignition temperature of various materials may estimate the temperature in a room by the degree of involvement of the materials affected. The ignition temperatures of some materials commonly found in buildings are given in Table 3.1.

TABLE 3.1
Ignition Temperatures of Common Combustibles

Dry wood (approximately)	500°F	(260°C)
Paper (newsprint)	451°F	(232°C)
Cotton cloth	439°F	(226°C)
Pyroxylin plastics	275°F	(135°C)

Construction materials and the contents of a building may be exposed to temperatures above their ignition temperature and still not burn, because of a lack of oxygen in the confined area or a short-duration exposure. The hazard lies in the fact that preheated combustible materials will burst into a free-burning fire when oxygen or air is admitted. Heating materials to their ignition temperature actually vaporizes the lighter fuel elements and compounds from the material into a gaseous form. These fuel gases combine with the gases of combustion and further increase the flammability of the atmosphere in the room. This fact is evident where, after a fire, charred woodwork is found where there was no flame impingement.

Consider backdraft or flashover possibilities

PERFORMING RESCUE WORK

When it becomes necessary to rescue occupants from imminent danger in a burning building, obviously this task should be given priority over all other activities. Ventilation and fire extinguishment may be carried on simultaneously with rescue procedures, but they are secondary in importance. Even if a building is filled with smoke and toxic gases, it may be necessary to enter for rescue work before it is thoroughly ventilated. The point of entrance should be determined to provide the greatest advantage to the rescuers and the least possible aid to the fire. Full protective clothing and breathing equipment are essential for rescue personnel.

To determine rescue procedures, consider again the three phases of burning as discussed in Chapter 2. In the first phase, there is a plentiful supply of oxygen present to support breathing and burning. In the second phase, the oxygen content of the room has been reduced, but enough remains to support breathing and burning. The upper level of the atmosphere in the building will have become hot, and a considerable amount of smoke and toxic gases will be present. The third phase consists of a fire that is smoldering, and the room is completely filled with smoke and noxious gases that are raised to extremely high temperatures. The oxygen supply is so deficient that rescuers without protective breathing equipment might be overcome in a comparatively short time.

Rescue procedures would necessarily be different during each phase of the fire. During the first phase, rescue might not be required at all. During the second phase, rescue and ventilation might be performed simultaneously. During the third phase, before the building is thoroughly ventilated, rescue procedures may become extremely difficult. This difficulty may be because of the lack of oxygen, obscured vision, the presence of toxic and noxious gases, high temperatures, and backdraft hazards.

Fire phase affects rescue

Rescue procedures include entering the building, locating the victims, releasing and removing the victims, and self-protection. Entering a building may be necessary by forcible entry, and, unless the fire has reached the third phase, entry may be made where desired. If the third phase has been reached, the point of entrance should be made only after ventilation has been accomplished.

ANTICIPATING THE SITUATION

Anticipation is a process of visualizing something that is not, at the moment, present before the eyes. When a fire officer has previously observed something or read something and later finds that it is necessary to visualize the situation under fire conditions, the officer is anticipating what is likely to happen. Anticipation must be based upon facts and existing conditions because a future course of action depends upon these two factors. The following sequence of anticipating is suggested, which is the first stage of size up:

Size up the situation

- Time of day.
- Occupancy.
- Probable location of the fire.
- Contents (probable fuel involved).
- Composition of fire gases.
- Type of building.
- Anticipated action of fires burning freely.
- Anticipated action of smoldering fires.

The need for previous inspection surveys and familiarization tours of buildings cannot be stressed too strongly. With such vital information, fire officers are in a position to anticipate situations. After an alarm has been received, a fire officer has only a few minutes to visualize and anticipate while en route to the fire. By the time the fire is in sight, the officer's basic thinking, using the knowledge gained through inspections, has been done. The officer is then ready for the second phase of size up: gathering the facts.

GATHERING THE FACTS

Facts concerning the situation are gathered after arrival at the fire. These facts have been previously considered in this manual as indicators of existing conditions. Since this is a study of ventilation practices during fire fighting, facts considered here will be confined to existing conditions that pertain to ventilation. Facts upon which ventilation procedures may be based include rescue, type of construction, and indications of flame, smoke,

smoke odor, and heat. These four last factors have been discussed separately and a consideration of them leads to the third phase of size up: evaluating the facts.

EVALUATING THE FACTS

The act of evaluating the facts is purely a mental procedure. The following list of conditions and requirements are not intended to be in the order of their importance, but the facts concerning them should be considered before any procedure or course of action is established.

Personal protection
Manpower and equipment
Rescue requirements
Occupancy
Location of the fire
Extent of the fire
Exposures
Extinguishment method
Building construction
Weather conditions
Material burning
Smoke and gas travel
Fire travel

DETERMINING THE PROCEDURES

With the situation anticipated, the facts gathered, and the facts considered, a fire officer can determine the procedures by which ventilation is to be carried out. The following sequence of procedures is recommended:

Ways to Ventilate
- Top/vertical: for example, opening roofs and skylights
- Cross/horizontal: for example, opening windows and doors
- Mechanical/forced: for example, smoke ejector and water fog

Rescue Procedures
- Establish need
- Use most desirable method

Fire Protection Requirements
- Exposures protected
- Charged lines in place
- Communications established

Attack and ventilation considered together

Ventilation Procedure
- Place(s) to ventilate
- Time to ventilate
- Methods of ventilation

Extinguishment Procedures
- Place(s) to enter
- Methods of extinguishing

TABLE 3.2
Ventilation Procedures

Hazards Involved in Ventilation	Possible Consequences	Approved Practice
Opening Below Fire	Backdraft No ventilation No value	Open above fire
Opening Too Soon	Lines not laid Fire gets away Increased loss Additional work time Additional life hazard	Lay and charge lines simultaneously Be prepared to fight the fire
Opening Wrong Place	No ventilation Backdraft Spread fire to area not involved Increased loss Additional work time Additional life hazard Adverse criticism	Have previous knowledge of building Open over vertical shafts Know contents Know hallways and partitions
Opening into Blind Attic (one that does not have access hole)	No ventilation Backdraft Loss of time Additional loss Additional life hazard	Have previous knowledge of building Use pike pole to open ceiling below
Insufficient Opening	Ineffective ventilation Backdraft Loss of time Additional loss Additional life hazard	Open large enough hole in first place Know size and location of areas below
Involvement of Exposures	Spread of fire Increased loss Additional work time Additional life hazard	Know surrounding buildings in advance Know horizontal openings Know roof structures Watch wind direction and velocity Have lines covering exposures Know heat and gases
Life Hazards to Firefighters	Roof collapse Explosions below Bad smoke and gas conditions Escape difficult in emergency	Paint top 18 inches of ladders (reflective white) Know building construction in advance If doubtful, lay ladder flat on roof

Hazards Involved in Ventilation	Possible Consequences	Approved Practice
	Injury from other's use of tools	Delay lower openings for ventilation above Know contents of building Cut from windward side of opening Place escape rope Two men cutting work from diagonally opposite corners of cut Have charged line at opening
Ventilation Delayed	Backdrafts Roof unsafe Entire building involved Excessive losses Increased life hazard	Prompt decisions Prompt action
Public Criticism	Hurts reputation of department Lowers morale of department Lessens public support Promotes political interference Lowers departmental efficiency	Know buildings in advance Know principles of ventilation Know proper methods of ventilation Use judgment Do a workmanlike job Explain department practices to public Explain reasons for ventilation Admit your mistakes Have department trained and disciplined Have knowledge of smoke and gases

VERTICAL (TOP) VENTILATION

NFPA STANDARD 1001

Fire Fighter I

3-10 Ventilation

3-10.2 The fire fighter shall identify the dangers present, and precautions to be taken in performing ventilation.

3-10.5 The fire fighter, using an axe, shall demonstrate the ventilation of a roof and a floor.

Fire Fighter II

4-10 Ventilation

4-10.1 The fire fighter shall demonstrate the use of different types of power saws and jack hammers.

4-10.2 The fire fighter shall identify the different types of roofs, demonstrate the techniques used to ventilate each type, and identify the necessary precautions.

4-10.3 The fire fighter shall identify the size and location of an opening for ventilation, and the precautions to be taken during ventilation.

4-10.4 The fire fighter shall demonstrate the removal of skylights, scuttle covers, and other covers on roof tops.

After an officer has sized up the situation for ventilation, determined the point at which the building should be opened, and provided adequate protection facilities, the next step is to open the building for ventilation. Although ventilation is not a method of fire extinguishment, it is a means by which firefighters can fight the fire more effectively and prevent property damage. A smoke-filled building does not necessarily mean that the gases are heated to the point that there is danger of a backdraft explosion, since the fire may be in the second phase and producing a great amount of smoke. The walls and contents also may have absorbed much of the heat from the gases. Under these conditions it may be possible, with the use of protective breathing equipment, to enter the building and extinguish the fire without ventilation, but releasing this contaminated atmosphere may simplify fire fighting and reduce damage by lowering its temperature and improving vision, which allow the firefighters to move faster and with more safety.

One common method by which smoke and heat may be removed from a burning building is by opening the structure at a strategic place to permit smoke and heat to escape. This method may be applied to top or vertical ventilation as well as to cross or horizontal ventilation. Although many situations and conditions commonly encountered in vertical ventilation are also common in horizontal ventilation, specific features are discussed here that influence top or vertical ventilation.

THE BUILDING INVOLVED

It has already been pointed out that knowledge of the building involved is a great asset when making decisions concerning ventilation, particularly when top or vertical ventilation is to be applied. Special construction features such as staircases, shafts, dumbwaiters, ducts, and roof openings are determining factors. Building permits that are issued in one's own city enable the fire department to know when buildings are altered or subdivided. Checking these permits will often reveal information concerning heating, ventilating and air conditioning systems, the type of roof construction, roof openings, and avenues of escape for smoke, heat and gases. Inspection and preplanning of all buildings, with emphasis on construction, is necessary for the application of proper ventilating procedures. The extent to which a building is connected to adjoining structures also has a bearing upon how top or vertical ventilation is undertaken.

Knowledge of building an asset

LOCATING THE FIRE

In most instances ventilation should not be carried out until the location of the fire is established. Smoke that is coming out of the top floor does not always indicate a fire on the top floor. The fire may be on a lower floor or even in the basement. Opening for ventilation purposes before the fire is located may spread the fire throughout areas of the building that would not otherwise have been affected.

Locating the fire

The fire may be located by feeling the wall, windows, or roof to determine heat conditions. Smoke that is gently flowing from an opening is not necessarily close to the seat of the fire. Obviously, extensive roof ventilation may be impractical or extremely dangerous if the location of the fire is such that vertical ventilation will draw the fire into parts of the building that are not involved. The location of the fire may be accurately determined by a proper system of size up.

The fire may have traveled some distance throughout a structure by the time fire fighting forces arrive, and consideration must be given to the extent of the fire as well as to its location (Figures 4.1 and 4.2). The severity and extent of the fire usually depends upon the kind of fuel, the time it has been burning, installed fire protection devices, and the degree of confinement of the fire. The phase to which the fire has progressed is a primary consideration in determining ventilation procedures. Some of the ways by which vertical extension occurs are as follow:

- Through stairwells, elevators, and shafts by direct flame contact or by convected air currents.
- Through partitions and walls and upward between the walls by flame contact and convected air currents.

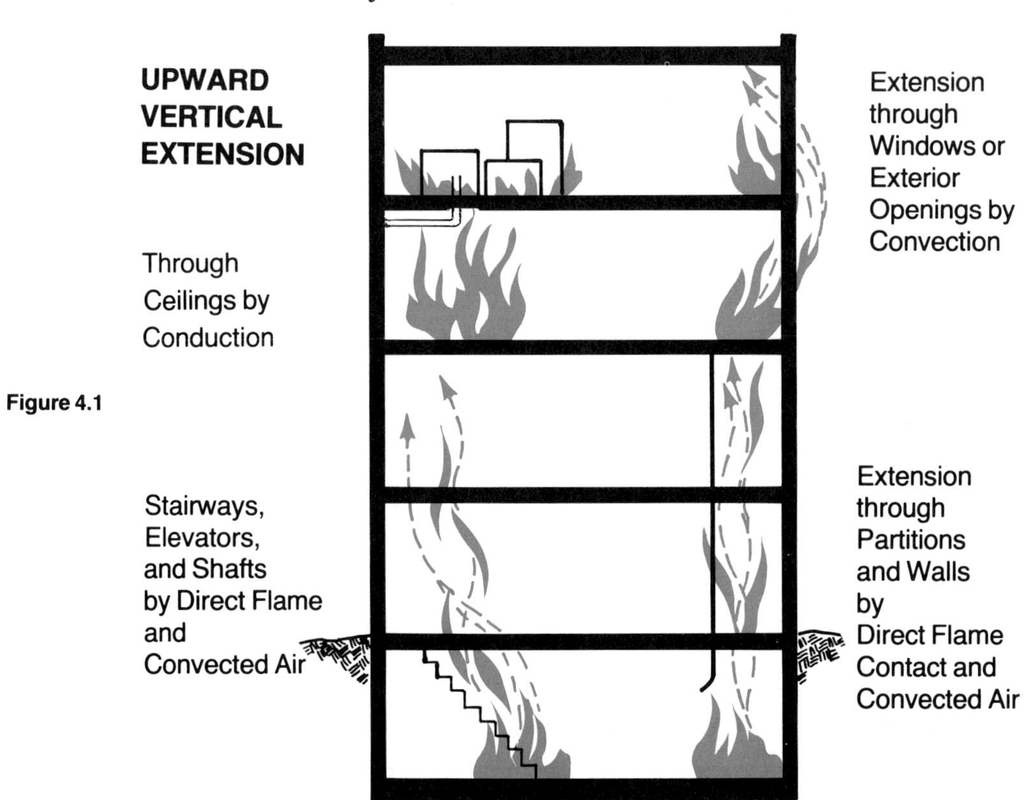

Figure 4.1

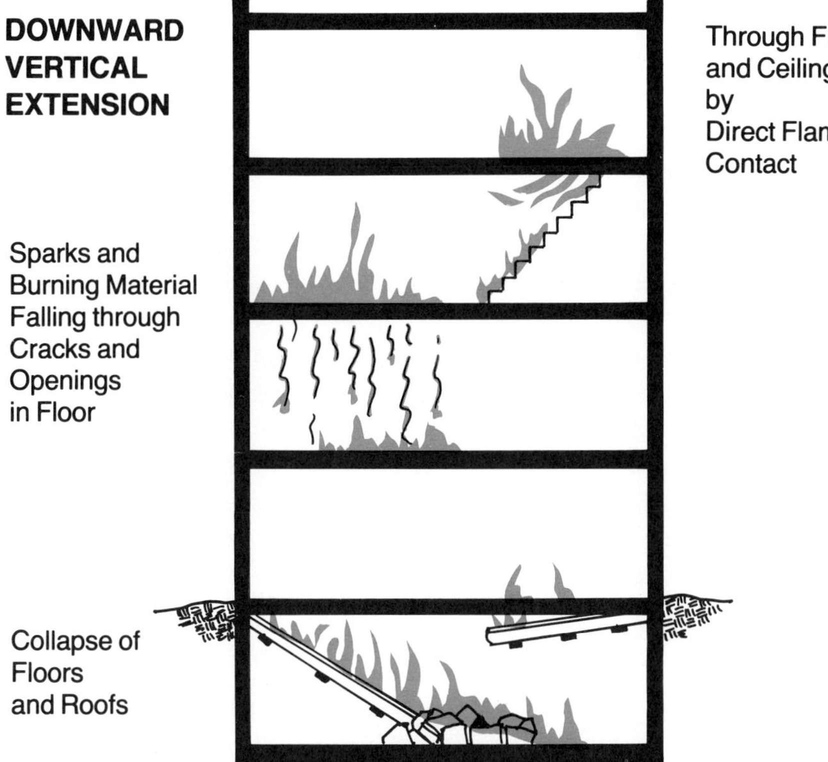

Figure 4.2

- Through windows or other outside openings where flame extends to other exterior openings and enters upper floors.
- Through ceilings, walls, and floors by conduction of heat through beams, pipes, or other objects that extend from floor to floor or room to room.
- Through ceilings and floors by direct flame contact.
- Through floor and ceiling openings where sparks and burning material fall through to lower floors.
- By the collapse of floors and roofs.

GETTING PERSONNEL AND TOOLS TO THE ROOF

In pre-fire planning, drills, and in actual fire attack, designated firefighters should automatically determine access to the roof. They are assigned the responsibility of accomplishing overhead ventilation when it is deemed necessary by the officer in charge. They may gain access to the roof by extension ladders, aerial ladders, or elevating platforms if these are not being used for rescue operations. A good standard procedure is to "spot" the aerial ladder at every fire and drill. Established as a standard operating procedure, this practice will save considerable time in ventilation and attack procedures. In those cases where the roof cannot be reached by fire department ladders, firefighters may use fire escapes or stairways within the building or in adjoining buildings if they can be used safely.

Under certain circumstances, practically every forcible entry tool can be applied to ventilation (Figure 4.3). In the opening of hatches and roof access doors, which may be locked, the axe and special prying bars and other tools may be required. Wire and bolt cutters will be necessary for removing locks and guy wires. Pike poles will be used for pushing away ceilings after the initial roof entry has been made. The opening may be made by the axe, by disc or chain saws, or by the use of oxyacetylene cutting equipment. The experienced firefighter will attempt to carry most of the tools that are thought necessary to the roof. If additional equipment is required, it will be acquired by either returning to the ground for it, hoisting it up by rope, or having another member of the crew deliver it. The care and use of these tools and the techniques of forcible entry are fully discussed in IFSTA 101, *Forcible Entry, Rope and Portable Extinguisher Practices,* and IFSTA 200, *Essentials of Fire Fighting.*

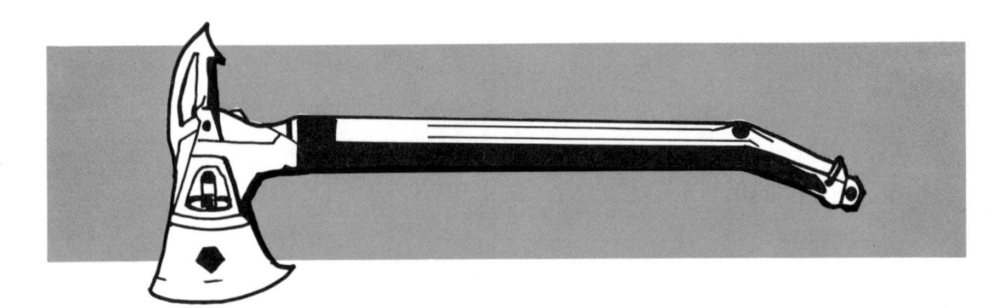

Figure 4.3 — An example of a multiple-use tool that can be used for ventilation.

USING NATURAL ROOF OPENINGS

Once firefighters have reached the roof, observed the condition of the fire (smoke), and the determination is made to open the roof, they should first look to natural openings for providing the necessary ventilation if other conditions favor such opening. Natural openings may exist in many forms. Scuttle hatches, skylights, monitors, ventilation shafts, and stairway doors are some of the more common openings provided that lend themselves to vertical ventilation. All of these may be expected to be locked or secured in some manner against entry.

Scuttle hatches are normally small, square, metal-covered hatches that provide entrance into the attic or cockloft (Figure 4.4 a, b). If the scuttle hatch is to be used, the ceiling directly beneath the hatch may require removal. Skylights should be removed, in preference to breaking the glass, if possible. Pry flashings loose on all sides and lift the skylight off completely (Figure 4.4 c, d). An alternative way would be to pry loose three sides and use the fourth side as a hinge. The sides of a monitor may contain glass (which is easily removed), and louvers of wood or metal. The sides that are hinged are easily forced at the top (Figure 4.4 e, f). If the top of the monitor is not removable, at least two sides should be opened to create the required draft. Stairway doors may be forced open in the same manner as other doors of like type.

Figure 4.4a and 4.4b — Scuttle hatches provide entrance to the attic or cockloft. They are usually locked on the inside with a padlock and will need to be forced.

Figure 4.4c — Skylights are good openings for ventilation. They should be removed, rather than having their glass broken.

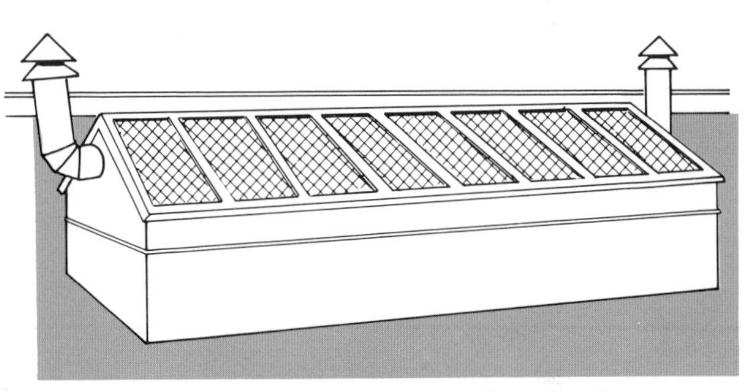

Figure 4.4d — Wired glass can be broken with the flat of an axe.

Figure 4.4e and 4.4f — The sides of a monitor should be easy to force. Open at least two sides for the required draft.

Roof obstructions can either help or hinder during ventilation operations. These obstructions may be air conditioning units, processing vents, machinery vents, dust collection units, and a variety of other items. A thorough analysis of all roof obstructions should be made while inspecting the structure. If it is determined that the venting unit or obstruction cannot be used for fire-related ventilation purposes, alternative methods of ventilation can then be devised and included in the pre-fire plan.

SELECTING THE PLACE TO VENTILATE

Open over the fire

The ideal situation in selecting a place to ventilate is one in which firefighters have prior knowledge of the building and its contents. There is no rule of thumb for selecting the exact point to open a roof except "as directly over the fire as possible." Many factors will have a bearing on where to ventilate. Some of them are:

- The availability of natural openings (skylights, ventilator shafts, monitors, hatches, and the like).
- Location of the fire and the direction in which the chief officer wishes it to be drawn.
- Type of construction.
- Wind direction.
- The extent the fire has progressed.
- The condition of the building and its contents.

The person accomplishing ventilation should also be constantly aware of the effect that ventilation will have on the fire and the consequence it will have on exposures. Prior to actually opening the building, consideration should be given to the department's state of readiness and its ability to protect exposures.

TOP-LEVEL VENTILATION PROCEDURE

Considerations for vertical ventilation

After the fire officer has considered the building involved and the location and extent of the fire, moved manpower and tools to the roof, observed safety precautions and selected the place to ventilate, the operation is not yet complete. Top-level ventilation involves all of these factors and many other precautions and procedures that the officer in charge must consider and practice if the top-level ventilation is to be successful. Prior to and during the actual opening of the roof, the officer in charge should give consideration to the following items:

- Coordinate with ground and attack companies.
- Observe the wind direction with relation to exposures.
- Note the existence of obstructions or weight on the roof.
- Secure a lifeline to the roof or provide another method as a secondary means of escape.
- Use natural roof openings whenever possible.
- Cut a large hole, if one is required, rather than several small ones (Figure 4.5).

Figure 4.5a — One large hole is better than several small holes. Cut as close to the ridgepole as possible.

Figure 4.5b — An improper and unnecessary ventilation job. Ventilation hole is not high enough — and fire was on other side of house.

- Find the fire location by looking for hot spots (melted ice or snow; sticky or bubbling tar).
- Imminent collapse of a flat roof may be indicated when the roof covering sags so joists can be seen as ridges.
- Ventilation of a flat roof may be done more safely and efficiently if the hole is cut three to five feet (1-1.5 m) from the hot spot.
- Exercise care in making the opening so main structural supports are not cut (Figure 4.6).
- Work with the wind at the back or side, to provide protection to the operators while cutting the roof opening.
- Guard the opening, to prevent personnel from falling into the building.
- Extend a blunt object through the opening, to break out the ceiling (Figure 4.7).

Figure 4.6 — For least structural damage, always sound for the location of rafters or trusses before cutting.

Figure 4.7 — After opening the roof, use a pike pole handle or similar blunt-ended tool to break out the ceiling. Make the ceiling opening at least as large as the roof opening.

SAFETY PRECAUTIONS

The ranking firefighter aloft should be in constant communication with the chief officer at the scene. Portable radios are most adaptable to this type of communication. Responsibility on the roof includes insuring that only openings are made that are required, coordinating the crew's efforts with those of the firefighters who are inside the building making the attack, and insuring the safety of all personnel who are assisting in opening the building. Some of the safety precautions that should be practiced include:

- Providing a secondary means of escape.
- Preventing personnel from walking on spongy roofs.

  This is usually a sign that structural members have been weakened. (If it is necessary to open a weakened roof, a roof ladder laid on the roof serves to distribute the firefighter's weight over a larger area.)
- Securing a lifeline to any firefighter who is to enter a weakened roof area.
- Protecting personnel from sliding and falling.
- Exercising caution in working around electric wires and guy wires.
- Insuring that the person making the opening is standing to the windward side of the cut and wearing the proper protective equipment.
- Not allowing other persons within range of the axe.
- Cautioning axe users to beware of overhead obstructions within the range of their axe.
- Starting power tools on the ground to insure operation at the site of the cut at upper areas. It is important that the tool be shut off before hoisting or raising the tool to these upper areas.
- Cautioning all cutting equipment operators to make sure the angle of the cut is not toward their bodies.
- Being on the lookout for indications of weakening structures or other hazards.
- Exercising caution, especially when using power tools, to cut or weaken supporting structural members.
- Keeping firm footing (Figure 4.8).

Figure 4.8 — Keep at least one foot on a roof ladder rung to prevent losing footing, especially on a steep or wet roof. (Note: Power saws should not be used on steeply pitched roofs, because of imbalance and danger of falling if the saw kicks.)

Potential hazards

Large flat or arched roofs covering undivided and unsupported areas present special hazards to firefighters. Single fire areas greater than 50,000 square feet are not uncommon. In some buildings open bays 40 by 50 feet with 20-foot or higher ceilings are required for the fabrication and storage of large items. The roofs of these structures, when supported by unprotected steelwork, may be expected to fail within a few minutes when exposed to intense heat (in many instances, less than ten minutes exposure causes collapse).

Large unbroken pitched roofs, which are normally used on barns, churches, and auditoriums, present peculiar ventilation problems. Firefighters frequently discover that it is not possible with standard equipment to reach the place where the opening should be made. In some instances, it is necessary to open this type of roof twice, once high up on the leeward side to provide an escape for the heated gases and again lower down to provide an intake for air. Ceilings under this type of roof are often located so far below the roof that they cannot be reached with the pike pole to open them.

TRENCHING

Trench ventilation, sometimes referred to as strip ventilation, is a form of ventilation that when properly done, will often prevent the horizontal spread of heat, smoke, and fire throughout a large building. This type of ventilation has been found to be very successful in controlling fires in buildings such as schools, apartment houses, motels, rest homes, and factories having long, open, undivided attic areas (Figure 4.9).

Figure 4.9a — Heat and heavy smoke issuing from trench cut, curtailing horizontal spread of fire. *Courtesy of Douglas Shelby.*

Figure 4.9b — Fire and smoke venting through trench cut, making area to the left more tenable for firefighters. *Courtesy of Douglas Shelby.*

Figure 4.9c — Interior lines are extinguishing fire and preventing horizontal extension. *Courtesy of Douglas Shelby.*

Figure 4.9d — View of the interior, showing fire side and tenable side. *Courtesy of Douglas Shelby.*

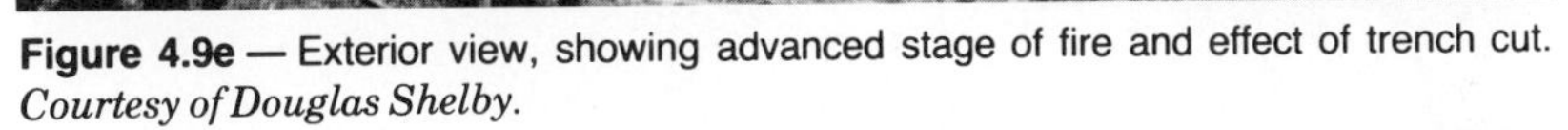

Figure 4.9e — Exterior view, showing advanced stage of fire and effect of trench cut. *Courtesy of Douglas Shelby.*

Figure 4.9f — The trench cut and interior attack confined the fire, saving much of the building. *Courtesy of Douglas Shelby.*

Trenching is done by making a three- to four-foot wide opening in the roof, then running this cut across the entire width of the building. Firefighters must remember to open the ceiling below once they have completed opening the roof. *It is imperative that the trench location be planned and sufficient time be allowed to permit completion of the opening before the fire reaches the trench.*

The opening, when completed, will allow a damper effect to be established, drawing heat and products of combustion out of the opening, preventing a horizontal spread of the fire. Interior hoselines should be placed in uninvolved areas below the trench cut and directed toward the fire area to aid in stopping the fire spread. Other streams can be used to protect the roof (Figure 4.10).

Many times flames will appear at the opening in the roof. *At no time should a hoseline or elevated stream be directed into this opening.* By directing water onto the flames, all of the heat and smoke will be forced back into the building.

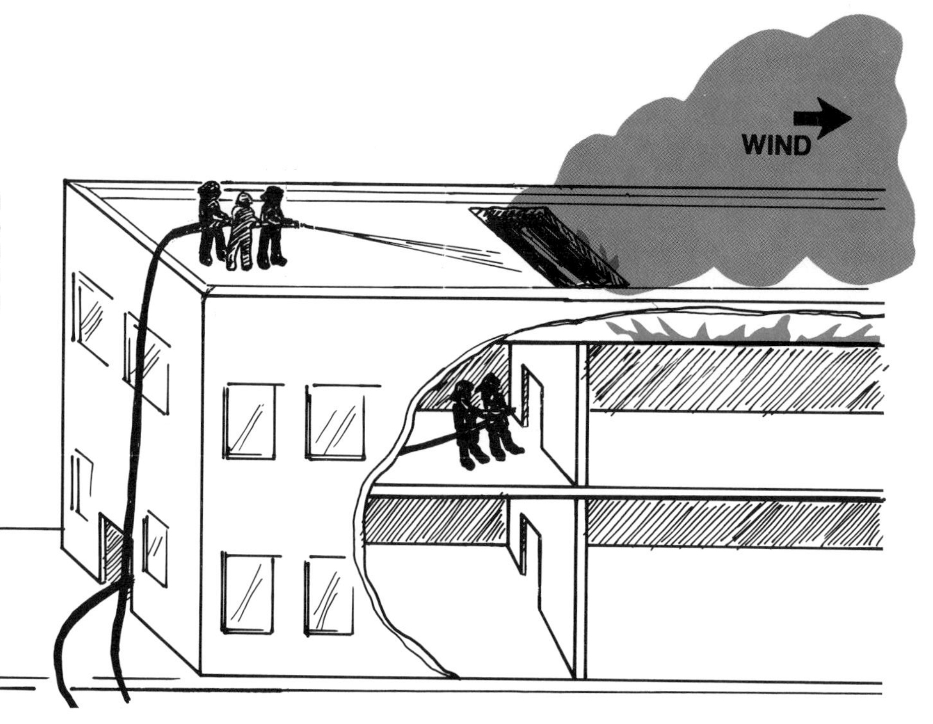

Figure 4.10 — Firefighters have completed the trench ventilation of this roof. Fire extension to the roof is being prevented by a stream that does not interfere with ventilation. Also, an interior line is being advanced to prevent horizontal extension.

OPENING THE ROOF

When the requirement to open the roof is recognized and natural openings do not exist or are not adequate to release the accumulated heat and smoke, the officer in charge may order the roof cut open (Figure 4.11). In some departments this decision is delegated to experienced officers. If the roof needs to be opened, the officer in charge should not hesitate. In making the decision, consideration must be given to the attack crews and the degree of assistance that this act will provide them. The initial opening should be made large enough to do the job, since the time required to make several smaller holes is greater than for one large one. For example, one 8 by 8-foot hole is equal to four 4 by 4-foot holes. When cutting through a roof, the firefighter should make the opening square or rectangular to facilitate repairs being made more easily.

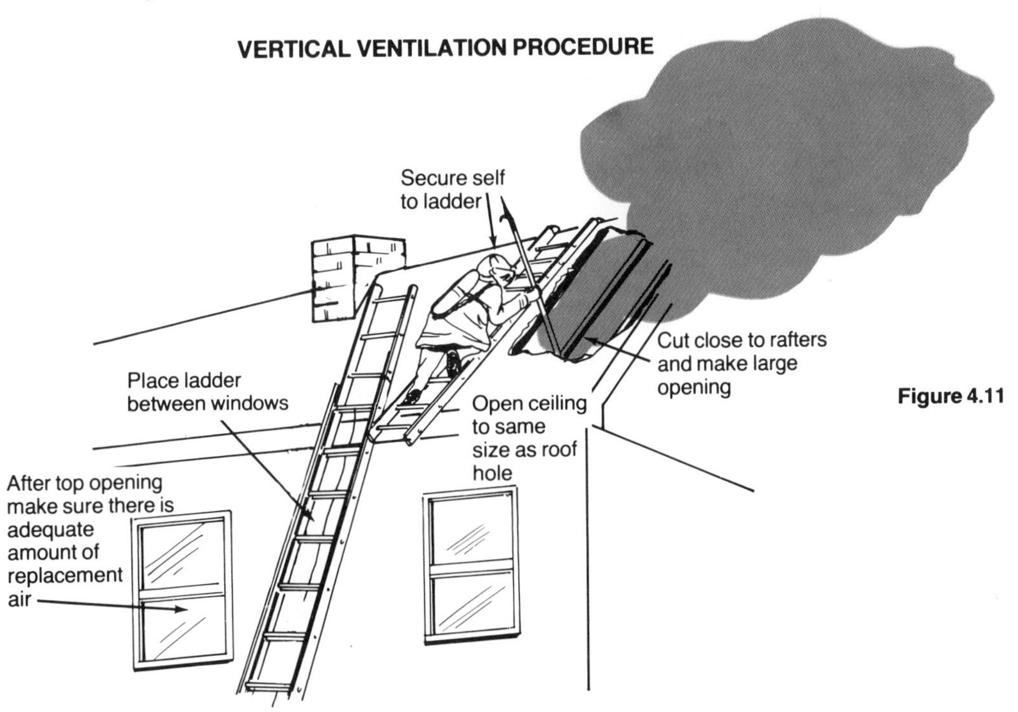

Figure 4.11

Pitched Roofs

Figure 4.12 — Gable roof.

Figure 4.13 — Hip roof.

Figure 4.14 — Modified gambrel roof.

Tools	Axe, power saw (general-purpose blade), pike pole, ladders, and lifeline.
Where	Laterally along ridge, leeward side first, over fire.
Construction	Wood and metal. Rafters or trusses generally run from eave to ridge. Sheathing placed on top, perpendicular to rafters or trusses.

Cut — Parallel to the rafters, or perpendicular to the ridge for approximately three feet. Pull sheathing or plywood with roof covering, using the pick of the axe or other tools. *Note:* On metal-covered roofs it should be possible to remove an entire section at one time. Pry along the edges, pulling screws or nails as necessary, and remove panel from leeward side of roof.

Flat, Sloped, and Modern Mansard Roofs of Wooden Construction

Figure 4.15 — Flat roof.

Figure 4.16 — Butterfly roof (a type of sloped roof)

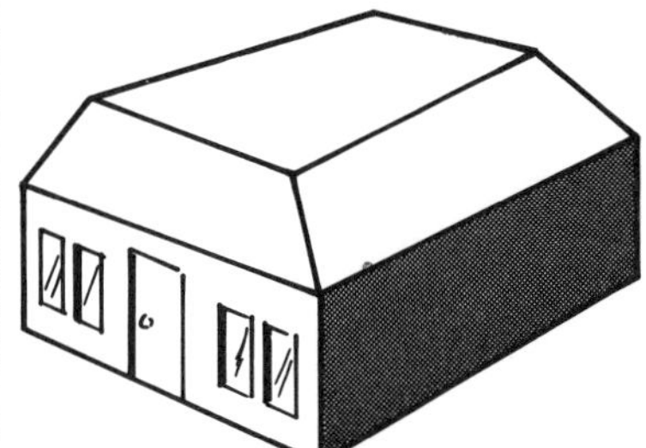

Figure 4.17 — Mansard roof

Tools — Axe, power saw (general-purpose blade), pike pole, stripping ladder, ladders and lifeline.

Where — Over seat of fire.

Construction — Wood. Rafters usually run parallel to the narrow dimension of the building. Sheathing is placed perpendicular to the rafters or narrow side.

Cut — Two cuts parallel to narrow dimension along rafters (might have to cut covering first). Make third cut perpendicular to narrow side of roof on leeward side. Open with pick of axe, insert stripping ladder or pike pole, and roll back sheathing and roof covering. Use butt of pike pole or stripping ladder to remove ceiling.

Metal Deck Flat Roof

Tools — Axe, power saw (general purpose blade for up to 18-gauge metal and metal blade for heavier), pike pole, ladders, and lifeline.

Where — Over seat of fire.

Construction — Metal bar joists usually run parallel to narrow dimension of building, metal decking is perpendicular to joists or narrow side. Built-up roof is placed over the metal panels.

Cut — Two cuts parallel to narrow dimension along joists. Third cut perpendicular to joists on leeward side. Avoid cutting joists, if possible. Pry up with the pick of the axe, breaking the tack welds. Roll back with stripping ladder or pike pole. Remove suspended ceiling below.

Figure 4.18a — Construction of a metal deck roof is shown in this photograph. Early collapse can be expected from exposure to heat and fire.

Figure 4.18b — Note the direction of the metal bar joists beneath the deck and supporting I-beams when installed.

Large Metal Deck

Tools	Same as metal deck.
Where	Same as metal deck.
Construction	Supporting beams run parallel to the narrow side of the building, and bar joists perpendicular to the beams and narrow side. Sheathing is parallel to narrow side and beams.
Cut	Two cuts perpendicular to the narrow side and beams, or parallel to bar joists. The third cut is parallel to the narrow side on the leeward side of the opening. Avoid cutting joists, if possible. Pry up with pick of axe, breaking tack welds, and roll back with pike pole or stripping ladder. Remove suspended ceiling below.

Concrete

Tools	Air hammer, power saw with concrete blade (if concrete is less than four inches thick).
Where	Over fire area or vertical shaft.

Construction	Poured concrete on reinforced concrete beams.
Cut	Generally it will not be possible to open roofs of this construction. Ventilate by using existing roof openings such as elevator or stairwell shafts, ventilators, or scuttle holes. Cross ventilation can also be used most effectively. Do not use building's mechanical ventilation systems.

Use natural openings

Pitched or Arched Trussed Roofs

Tools	Axe, power saw (general-purpose blade), pike pole, stripping ladder, ladders, and lifeline.
Where	At top of bow or peak over fire, or long narrow cut at top of bow or peak over several rooms.
Construction	Wood or metal. Trusses placed across narrow side of building or multiple bays of trusses. Rafters run perpendicular to trusses. Sheathing runs parallel to trusses.
Cut	Two cuts perpendicular to trusses or narrow side of building, along rafters. Make third cut parallel to truss on leeward side. Open with pick of axe, insert stripping ladder or pike pole, and roll back sheathing and roof covering. Use heel of pike pole or stripping ladder to remove suspended ceiling if necessary.

Tile/Slate Roofs

Tile roofs, such as found on Spanish architecture, or slate roofs, often found on churches or other large buildings, present special problems in ventilation (Figure 4.19). A tile roof is made of curved tiles that are "nested" on a roof, usually over a layer of tar paper and wooden sheathing. The tile/slate is extremely fragile and usually cannot be walked on without breaking. This creates two problems for ventilation crews trying to reach the highest point of the roof. First, the crew will do a lot of damage getting to the desired point of ventilation. Second, the broken tiles or slate make a very difficult walking surface. These pieces are also very heavy and may slide off the roof, creating a safety hazard for those below.

Tile/slate roof hazards

When a tile/slate roof has to be ventilated, a roof ladder should be used to give the personnel good footing. The best tool to start tile/slate roof ventilation is a sledge hammer, which can shatter the tiles or slate over the designated area better than a pick-head axe can. The cutting of the sheathing may be accomplished using an axe or a power tool. *Caution: Tile/slate roofs carry more weight per surface area than any other roof style and are very susceptible to early roof collapse.* Roofs with this much weight also become "spongy" earlier than one might expect. Because of this weight factor, ventilation crews must always be sure to coordinate their efforts with attack crews and, especially, warn them of the potential of falling debris. A "spongy" roof is one that has been weakened by fire. Indications of a spongy roof include sagging under a person's weight and lack of resistance to an axe (the wood will act punky). Many modern roofs are "springy"; that is, they will give some when a person walks on them at any time. Pre-fire inspections should include checking to see whether roofs are springy. This kind of information could save lives and the structure in the event of a fire.

Figure 4.19a — Tile roof.

Figure 4.19b — Ventilating a tile roof.

CHURCH FIRES

When fighting fires within churches, vertical ventilation becomes almost impossible because of the high, steep roofs that many churches have. Many times the height of the roof is greater than the length of the roof ladders carried by the fire department. Often, because of the height of the building and the angle of the roof, aerial ladders must be used. Even if vertical ventilation is possible, many old churches have a dead air space of 12 to 18 feet between the ridge of the roof and the ceiling below. It is highly unlikely that the average fire department will have a pike pole long enough to remove the ceiling below through the roof opening. Many old churches have ceilings that are constructed of heavy plaster. During a fire, these ceilings may collapse and injure firefighters below.

Horizontal ventilation may be the only possible way in which ventilation may be accomplished. Even this will be difficult because many modern churches are built windowless or with very few windows at all. In older churches, windows may be made of stained glass, which is very costly to replace and may have much sentimental value to the community. Firefighters should not hesitate to break out stained-glass windows if it is the only way to provide some type of ventilation to the structure. It is better to replace the stained-glass window than to replace the entire church.

PRECAUTIONS AGAINST UPSETTING ESTABLISHED VERTICAL VENTILATION

Firefighters at some time or another may have had the experience of being driven from a building by intense heat and smoke after action had been taken to ventilate the building. A condition such as this could be very surprising, especially if good, sound ventilation practices had been followed. Closer inspection may reveal that some factor was introduced that upset the established ventilation and its effect.

When vertical ventilation is accomplished, the natural convection of the heated gases creates upward currents that draw the fire and heat in the direction of the upper opening. Fire fighting teams take advantage of the improved visibility and less contaminated atmosphere to attack the fire at its lowest point, following through to extinguishment, roughly along the same route the smoke is traveling. If the "stack effect" is interrupted, the heat, smoke, and steam backs up, hampering extinguishment efforts. Some common factors that can destroy the effectiveness of ventilation are as follow:

- Improper use of forced ventilation.
- Breaking glass at wrong place or time.
- Improperly directed fire stream (Figure 4.20).
- Breaking skylights at wrong place or time.
- An explosion.
- A burn-through.
- Additional openings between the attack team and the upper openings.

Be careful with elevated streams

Elevated streams are frequently used to cut down sparks and flying brands from a burning building or to reduce the thermal column of heat over a building. When elevated streams are projected downward through a ventilation opening or used to reduce the thermal column to a point where ventilation is hindered, they either destroy or upset the orderly movement of fire gases from the building. An upset of this nature can materially affect firefighters who may be working at various levels on floors below. Elevated streams that are being operated closely above ventilated openings should be projected slightly above the horizontal. In this position they will help cool the thermal column and extinguish sparks. The movement of the stream may even increase the rate of ventilation.

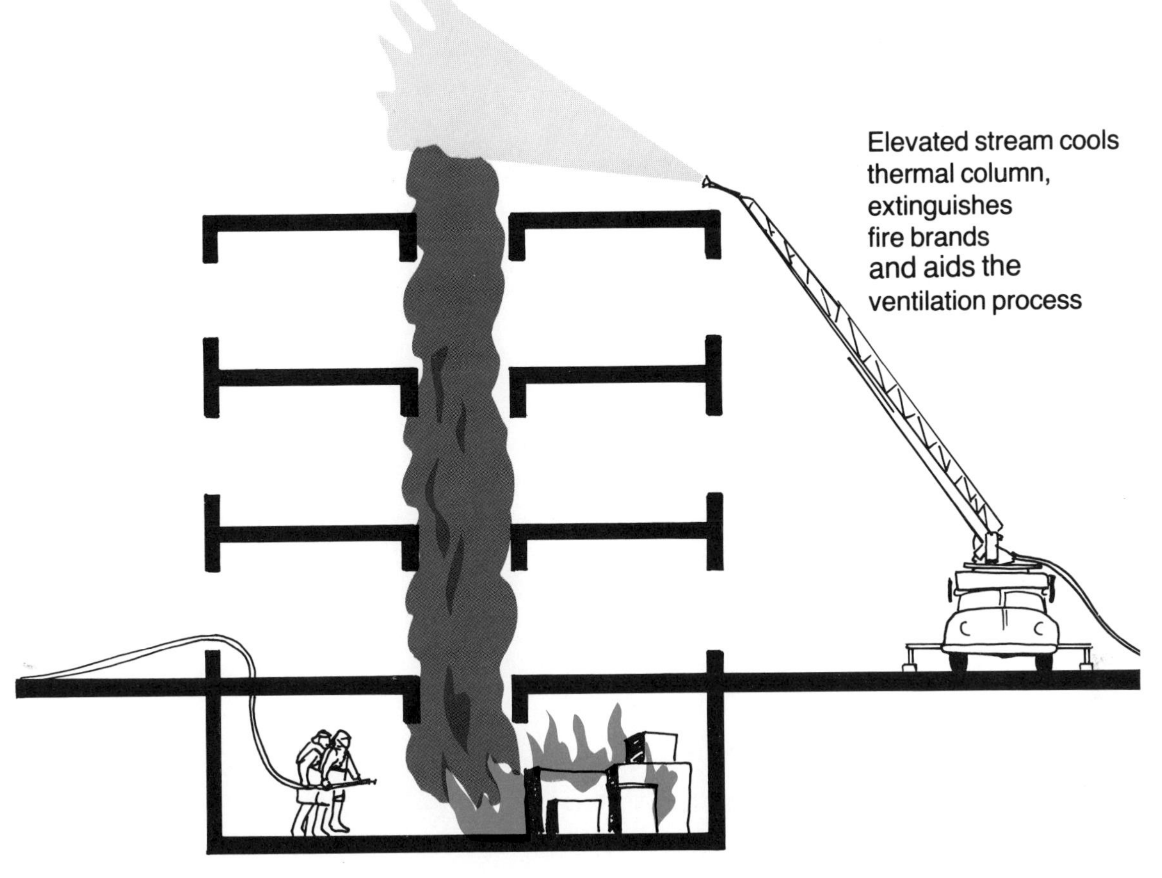

VENTILATION AND ELEVATED STREAMS

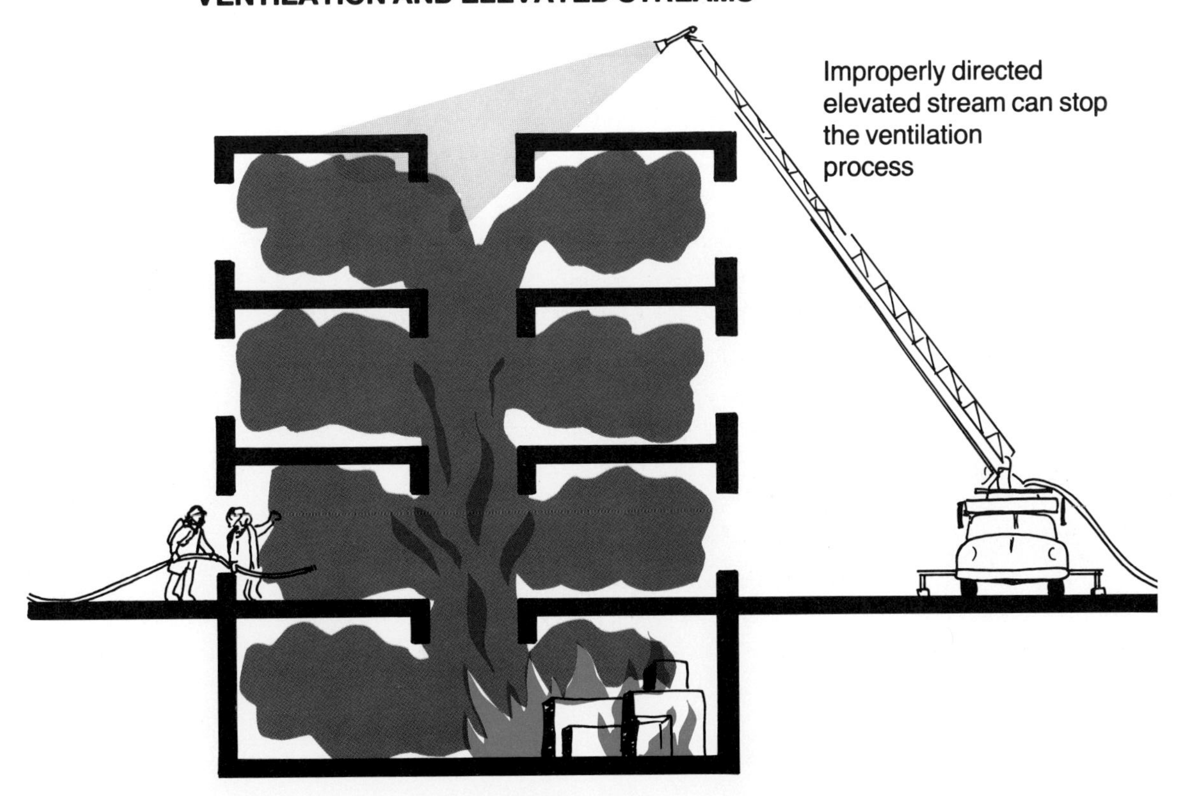

Figure 4.20

HORIZONTAL (CROSS) VENTILATION

NFPA STANDARD 1001

Fire Fighter I

3-10 Ventilation

3-10.3 The fire fighter shall demonstrate opening various types of windows from inside and outside, with and without the use of fire department tools.

3-10.4 The fire fighter shall demonstrate breaking window or door glass, and removing obstruction.

Thus far, only vertical ventilation has been stressed. Although this method may be fundamentally typical in fire ventilation, vertical ventilation is not the solution to all ventilation problems. There may be many instances where its application would be impractical or impossible because of conditions and considerations previously discussed. Only prompt and accurate size up based on a thorough understanding of the many variables that affect and are affected by ventilation can provide the answer to the following two questions:

- Is ventilation required?
- What method is most appropriate in this instance?

Horizontal ventilation is a method that may prove to be more effective for a particular situation or it may be used in conjunction with other methods. In addition to selecting the method of ventilation, a fire officer must determine how, when, and where it will serve its most useful purpose. A discussion of some of the factors that influence horizontal ventilation follows.

STRUCTURAL CHARACTERISTICS OF THE BUILDING

Determining need for horizontal ventilation

Building type, design, and occupancy are some of the initial factors that must be considered in determining whether either horizontal or vertical ventilation should be accomplished. Type and design features that have a bearing on the decision include the number and size of wall openings, the number of stories, the availability and involvement of exterior fire escapes, and exposures. Directly related to these are protective coverings of the openings, the direction in which the openings face, and the wind direction.

Both personnel and materials must be included in occupancy considerations. If preapproach information has been made available concerning the status of the people who occupy the structure, and if rescue is not required, the next consideration is to determine the presence of hazardous materials or processes that may endanger fire fighting crews. As has been previously emphasized, it is not possible to analyze the degree of toxicity of smoke under fire conditions; therefore, pre-fire planning should include familiarity with the type of materials involved in each major building under the jurisdiction of the department.

Structures to which horizontal ventilation is normally applied include:

- Residential buildings in which the fire has not involved the attic area.
- Buildings with windows high up the wall, near the eaves.
- The attics of residential buildings that have louvered vents in the walls.

- The involved floors of multistoried structures.
- Buildings with large, unsupported open spaces under the roof, in which the fire is not contained by fire curtains and in which the structure has been weakened by the effects of burning. (Generally, if a fire has been burning for 20 minutes, the roof can be considered unsafe for the weight of a firefighter.)

LOCATION AND EXTENT OF THE FIRE

Many of the aspects of vertical ventilation may also apply to horizontal ventilation. A different procedure, however, must sometimes be followed in ventilating a room, a floor, a cockloft, an attic, or a basement. The procedure to be followed will be influenced by the location and extent of the fire.

Some of the ways by which horizontal extension occurs (Figure 5.1) are as follow:

- Through wall openings by direct flame contact or by convected air.
- Through corridors, halls, or passageways by convected air currents, radiation, and flame contact.
- Through open space by radiated heat or by convected air currents.

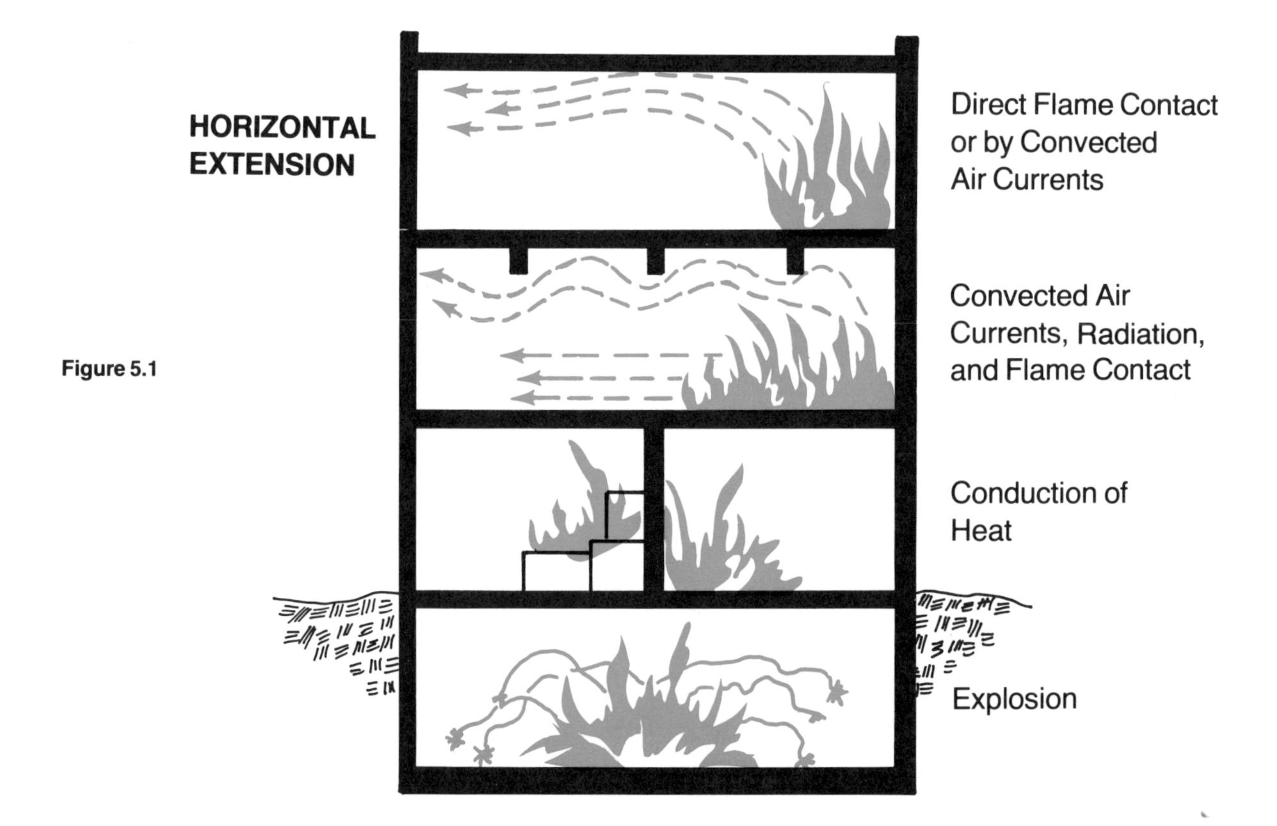

Figure 5.1

- In all directions by explosion or flash burning of fire gases, flammable vapors, or dust.
- Through walls and interior partitions by direct flame contact.
- Through walls by conduction of heat through beams, pipes, or other objects.
- Through air conditioning and heating ducts.
- Through plenums.

BREACHING FIREWALLS

During ventilation operations, firefighters must be careful not to breach or make openings in firewalls. As the ventilation process occurs, heat, smoke, and fire gases travel through openings that may also draw the fire toward the opening because of drafts produced and the availability of oxygen at the opening and the other side of the wall. Therefore, when a firewall is breached to make a ventilation opening, there is also the possibility of drawing heat, smoke, and gases through the firewall and endangering the previously protected portions of the building that the firewall was designed to protect.

Keep firewalls intact

The potential for flammable fire gases to accumulate on the uninvolved side of the firewall at the roof level is perhaps the most serious. These gases may accumulate, unknown to the firefighter, until a source of ignition reaches the vapors and ignites them, which may cause either a flashover or backdraft.

EXPOSURES

Since horizontal ventilation does not normally release heat and smoke directly above the fire, some routing is necessary. Firefighters, therefore, should be aware of internal exposures as well as external exposures. The routes by which the smoke and heated gases would travel to the exit may be the same corridors and passageways the occupants will be using for evacuation.

Consider exposures

Outside exposures include all those previously mentioned in the discussion of vertical ventilation, plus those that are peculiar to horizontal ventilation. Since horizontal ventilation is accomplished at a point other than at the highest point of a building, there is the constant danger that the rising heated gases will ignite the portion of the structure that they contact when released. They may ignite eaves of adjacent structures or be drawn into windows above their liberation point. Smoke may also be drawn by air conditioners into adjacent buildings (Figure 5.2).

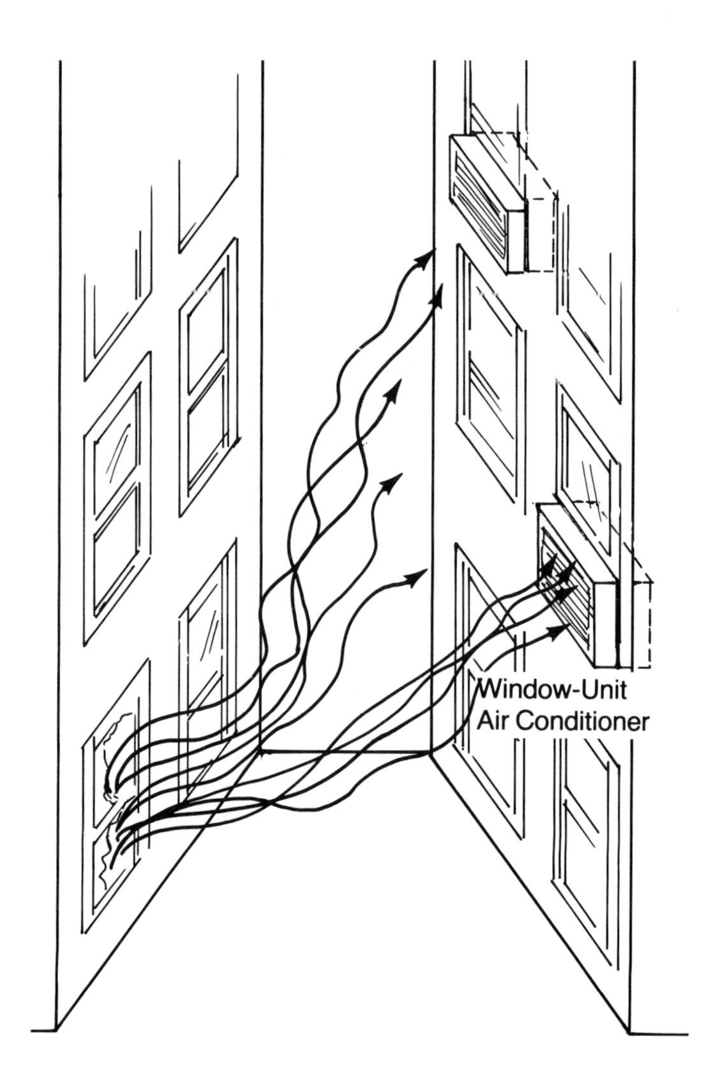

Figure 5.2 — Consideration must be given to where the smoke goes after it leaves the building.

WEATHER CONDITIONS

Wind

Weather conditions are always a primary consideration in determining the proper ventilation procedure. The wind plays an important role in ventilation. Its direction may be designated as windward or leeward. The side of the building where the wind is striking is windward; the opposite side is leeward. Under certain circumstances, when there is no wind, horizontal ventilation is not very effective, since the force that removes the smoke is absent. In other instances horizontal ventilation cannot be accomplished because of the danger of wind blowing toward an exposure or feeding oxygen to the fire.

Humidity

Allowances should be made during times of high humidity because ventilation is somewhat hindered. High humidity tends to keep the products of combustion from rising into the atmosphere. This tends to keep the products of combustion at or near ground level.

Inversions

Temperature inversions called *night inversions* also affect smoke behavior. During a night inversion, there is a layer of cool air at and above ground level. The air temperature increases with height. This temperature difference might be as much as 25°F in 250 feet. Smoke will rise until its temperature matches that of the air and will then extend horizontally (Figure 5.3).

Figure 5.3 — When there is a "night inversion" smoke will rise until its temperature equals that of the surrounding air, then the smoke spreads in a layer.

USING WINDOWS

The correct procedure for ventilating a building horizontally is to open the top windows on the leeward side and, after allowing the superheated gases to escape, open the lower windows on the windward side (Figure 5.4). When a window is opened for the purpose of ventilation, the screen, curtains, drapes, and blinds should be removed since they also serve to block circulation.

Figure 5.4 — Open upper windows on the leeward side first, allowing heated gases to escape, then provide windward openings as needed.

Breaking Windows

Always try to open a window before assuming it is locked. If it is necessary to break a window, the entire pane should be broken out and the sash cleared of broken glass by scraping the sash with the breaking tool to remove all broken pieces. This procedure serves a two-fold purpose. First, if it is necessary to break out the window, maximum effect should be obtained. Second, injuries will be prevented to those who may come into contact with the sash. The firefighter who breaks out the window should do so with the flat side of the axe blade or other tool while standing to the side of the window. The handle of the tool should be held higher than the blade, to prevent glass from sliding down the handle (Figures 5.5 and 5.6).

The firefighter who breaks out windows from a ladder should use the pike pole to open top windows first. This method will release smoke and heat from the interior of the building better than from lower windows, which would obscure the lower work area and endanger the firefighter. The firefighter should not be positioned directly above or downwind from the window to be opened and should stay clear of the path of the escaping smoke and heat. Someone should be positioned on the ground to warn crews that falling glass is expected in the area.

If it is imperative for personal safety to break window panes and no forcible entry tools are immediately available, a firefighter's helmet may be used if safety precautions are followed to protect exposed body surfaces. A better solution is to carry suitable hand tools.

Figure 5.5 — Look below before breaking glass, warn others, then break glass safely.

Figure 5.6 — With some forethought, breaking windows can be facilitated. When possible, break windows from the roof or from central locations. A fire stream directed from the ground may also be used to break windows if there is not other means. This illustration is a composite of several possible methods.

Thermoplastic Windows

A current trend in the construction industry is the use of Plexiglas and other thermoplastics in place of glass windows. Lexan is an example of a polycarbonate that has seen wide application as a glass substitute for its ability to withstand abuse from vandalism or weather. Lexan is 250 times stronger than safety glass, 30 times stronger than acrylics, and is classified as self-extinguishing. It is 50 percent lighter than glass and 43 percent lighter than aluminum. Lexan is available in thicknesses ranging from ⅛ to 1 inch, with the most common thicknesses being ⅛ to ½ inch. Various tests have been conducted by departments across the country to gain information on the best way of forcing entry into Lexan windows. The tests have used standard forcible entry equipment including pick-head axes, oxyacetylene cutting torches, reciprocating saws, and circular saws. These tests, along with actual field experiences with Lexan during fires, indicate that a circular saw with a carbide-tipped blade is most effective when entry must be made through Lexan. Caution must be taken in blade selection since a blade with too fine a tooth will melt the Lexan and cause the blade to bind. Conversely, a blade that is too coarse will cause the blade to slide dangerously over the cutting surface. A medium-tooth blade (approximately 40 teeth) has been found to yield the best results.

If a power saw is not available, entry might be made in one of two other ways. One is to strike the pane in the center with a sledge hammer. This will not break the pane, but will bow it so it is likely to slip out of its keepers. Another method is to use the point of a pick-point axe to score an X on the pane, then strike the pane. The pane will usually break along the X and the pieces can then be pulled or bent out. Neither technique will work if the pane has been heated to the point where it becomes somewhat soft and pliable.

Lexan windows difficult to break

It must be reported that during actual fire conditions, even where companies had been made aware of the Lexan characteristics, much difficulty was encountered in ventilating buildings that had Lexan windows.

OBSTRUCTIONS TO HORIZONTAL VENTILATION

Buildings that contain a large number of rooms may be very difficult to ventilate horizontally because of the absence of air circulation between the rooms. The building may be completely filled with smoke that accumulated over a period of time and the smoke may be filtering through cracks and ventilator openings. If ventilation is attempted without a knowledge of the location of the fire or the arrangement of rooms and partitions in the building, uninvolved sections of the building may be jeopardized. In addition to walls and partitions, stacks of stored material may obstruct horizontal ventilation. See Chapter 6 for more information concerning obstruction.

PRECAUTIONS AGAINST UPSETTING ESTABLISHED HORIZONTAL VENTILATION

Personnel should also be careful not to block or close openings that channel fresh air into the area that is being ventilated. Established ventilation may also be upset if additional openings are made that rechannel the air currents that are supposed to be ventilating the area (Figure 5.7).

Other hindrances of good horizontal ventilation are:

- Improper use of forced ventilation.
- Improperly directed fire streams
- Improper placement of salvaged contents with salvage covers covering items and obstructing air currents.

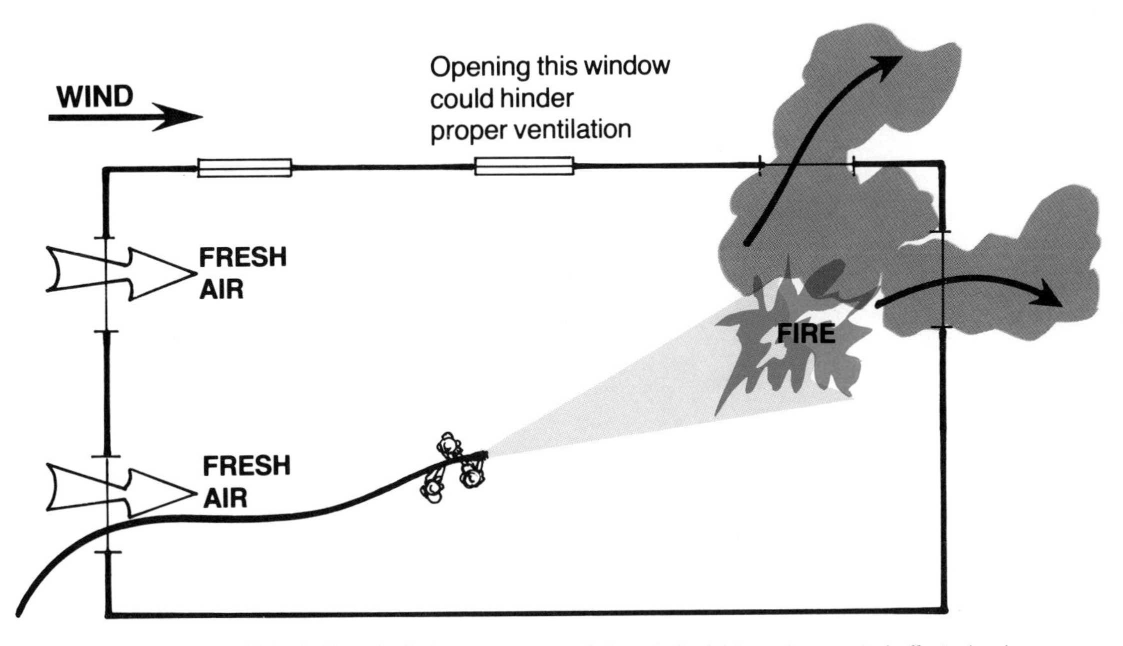

Figure 5.7 — If the indicated window were opened, the wind might create a venturi effect, drawing through that window both fresh air (from the openings on the left) and smoke from the fire. Even if the venturi effect were not very strong, opening the window would upset the relative positive pressures, hindering effective ventilation.

FORCED VENTILATION

NFPA STANDARD 1001

Fire Fighter II

4-10 Ventilation

4-10.5 The fire fighter shall demonstrate types of equipment used for forced ventilation.

4-10.6 The fire fighter shall demonstrate ventilation using water fog.

Ventilation has thus far been considered from the standpoints of the natural flow of air currents and the currents created by fire. Forced ventilation is accomplished by mechanical blowers, fans, or fog streams. The fact that forced ventilation is effective and can be depended upon for smoke removal when other methods are not adequate proves its value and importance.

SITUATIONS REQUIRING FORCED VENTILATION

There is no set rule for determining just when to use forced ventilation other than "when necessary for the removal of an undesirable atmosphere." In many instances different methods of ventilation are successfully applied in combination. Some of the situations that indicate the need for forced ventilation are listed below:

- When the type of construction is unsuitable for natural ventilation because of the lack of vertical shafts or horizontal exterior openings.
- When fire is below ground level in basements or below deck on marine craft where it is difficult to gain entrance.
- When there is no fire and it is necessary to remove undesirable atmosphere from areas such as tunnels, buildings, basements, and other enclosures.
- To relay smoke through buildings of large area.

Steps for successful forced ventilation

Basements frequently have inadequate ventilation because of the lack of windows, weather conditions, tight construction, and devices for protection against intruders. These features preclude natural ventilation and generally make forced ventilation necessary. Ventilation in basements and subbasements is normally difficult, presenting problems that are unique to this type of structure. In order for mechanical ventilation to be successful, the following tasks must be accomplished:

- Start ventilation as soon as the hoselines are attacking the fire.
- Place the ejector at an outside opening that is closest to the fire.
- Place the ejector as high as possible, for maximum effectiveness (Figure 6.1).
- Use the wind to your advantage to help the smoke ejector either push or pull out the heat and smoke.
- Place the ejector at the proper distance from the opening to prevent recirculation of the smoke (Figure 6.2).
- Remove items such as draperies, which might interfere with the airflow out of the ejector, from around the smoke ejector (Figure 6.3).

Figure 6.1 — Place smoke ejector as high as is practicable, to evacuate the risen hot air, gases, and smoke as quickly as possible.

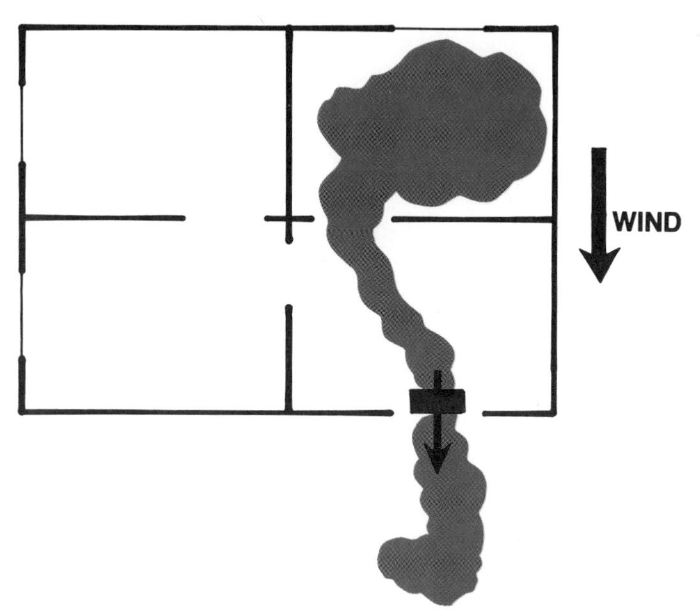

Figure 6.2 — When possible, direct forced ventilation in the same direction as the wind.

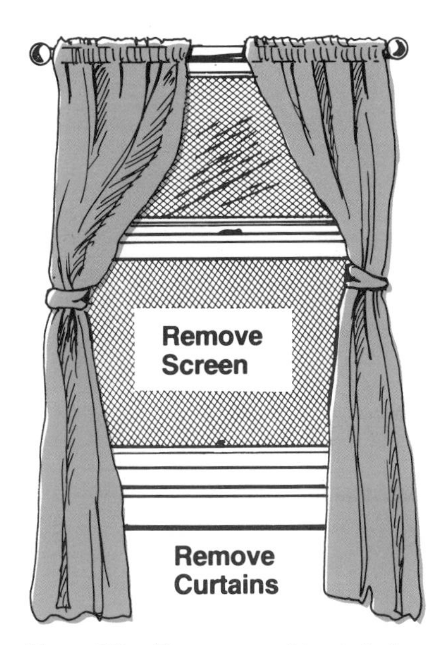

Figure 6.3 — Remove everything, including common window screens, that could impede free air flow.

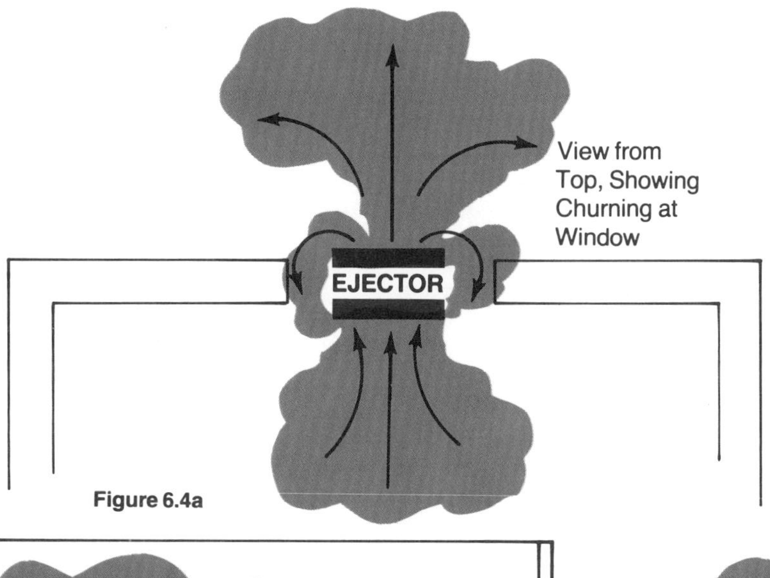

Figure 6.4a, 6.4b, and 6.4c — When using forced ventilation, prevent "churning" by blocking spaces around ejection mechanism. Use any suitable material available such as salvage covers, pillows, blankets, and curtains.

Figure 6.4b

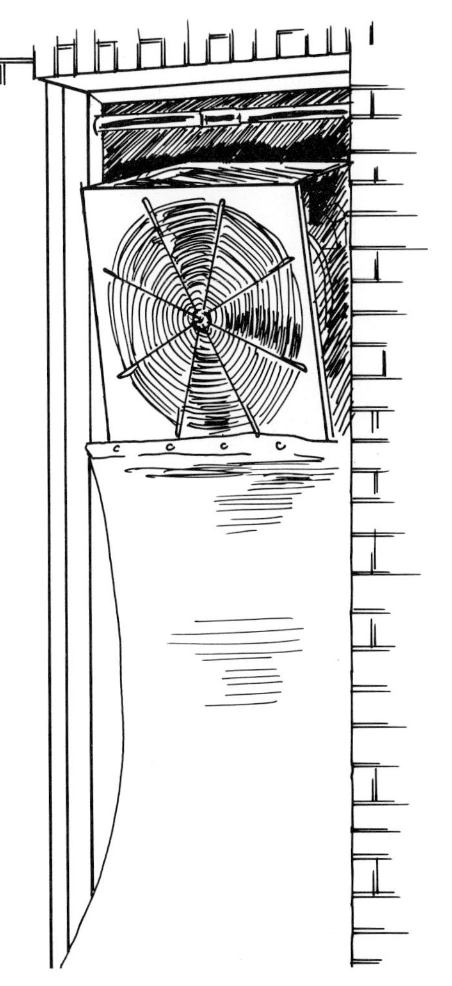

Figure 6.4c

- Use any available means to prevent "churning." (Figure 6.4.)
- Provide and maintain unobstructed replacement-air opening.

SOME ADVANTAGES OF FORCED VENTILATION

The value of mechanical ventilation may be further realized when in order to protect human life it becomes necessary to rapidly rid premises or areas of an undesirable atmosphere. Even though fire may not be a factor, contaminated atmospheres must be thoroughly ejected. In addition, mechanical ventilation can be reversed and used to introduce fresh air into work areas where firefighters or rescue personnel are working for prolonged periods of time in such activities as overhauling or salvage or are working in underground areas. When used in this manner, care must be exercised to insure that the ventilating units do not draw in contaminants. Forced ventilation, if not the only means of clearing a contaminated atmosphere, is always a welcome addition to normal ventilation. Some of the reasons for employing mechanical or forced ventilation are indicated in the following list:

- It insures more positive control of air movement.
- It supplements natural ventilation.
- It speeds the removal of contaminants, facilitating more rapid rescue under safer conditions.
- It may be used where other methods fail.
- It reduces smoke damage.
- It promotes good public relations.

SOME DISADVANTAGES OF FORCED VENTILATION

If mechanical or forced ventilation is misapplied or uncontrolled, it can cause a great deal of harm. Forced ventilation requires considerable supervision because of the mechanical force that is behind it. Some of the disadvantages of forced ventilation are indicated in the following list:

- It can move fire along with the smoke and extend it to lateral areas.
- It can introduce air in such great volumes that it can be the cause of a fire spreading.
- It is dependent upon power, the interruption of which renders it ineffective.
- It employs additional personnel for its operation.
- It requires special equipment.

FORCED VENTILATION EQUIPMENT

If natural or vertical ventilation is even partially unsatisfactory, mechanical ventilation should be employed (Figures 6.5-6.7). By using smoke ejectors to pull products of combustion away from occupants and their means of egress, an avenue of escape or rescue is provided.

Figure 6.5a — A typical electric-motor-driven smoke ejector. *Controlled Airstreams*

Figure 6.5b — An explosion-proof electric-motor-driven smoke ejector.

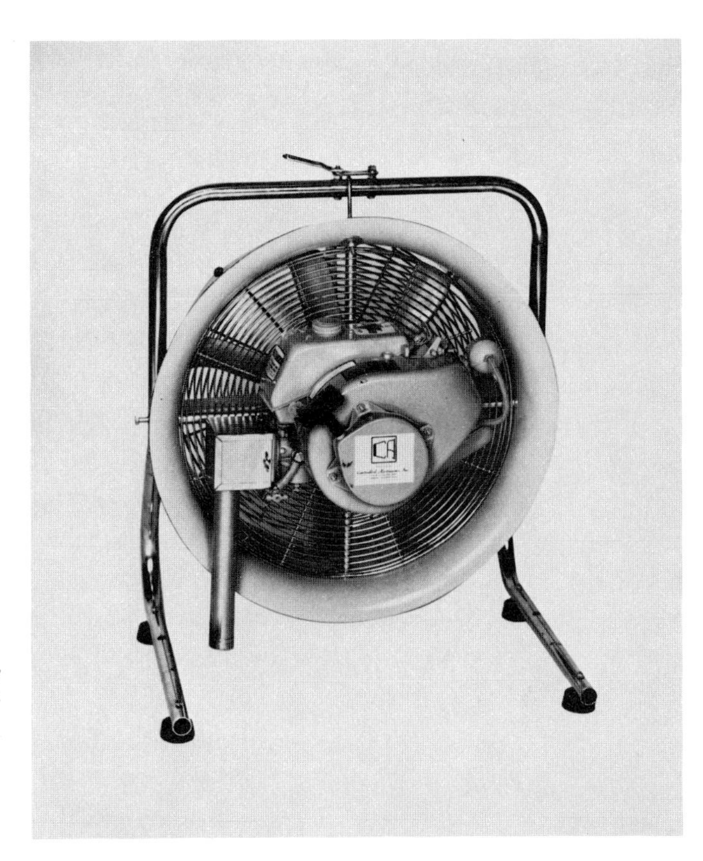

Figure 6.6 — Gasoline-engine-driven smoke ejectors. *Courtesy of Super Vacuum Mfg. Co. (above) and Controlled Airstream (right).*

Figure 6.7 — Water-powered smoke ejector. *Angus Fire Armour*

How to Achieve Maximum Benefits Through Mechanical Ventilation

- Use good equipment. Be sure to use a large enough ejector, or enough ejectors to perform the necessary task.
- Provide proper training in technique to acquire maximum benefit from the equipment.
- Practice good pre-fire planning. Know proper placement of equipment under any condition. Plan so equipment, training, methods, and tactics will work to accomplish the desired results with no time wasted.

Two primary functions of ejectors

Smoke ejectors perform two primary functions: They exhaust or pull the initial heat and smoke from a burning building, and, once the ejection action has started and the smoke becomes thin and less buoyant, they can also be used to blow the smoke from the building. A blowing ejector is also effective in eliminating heavier-than-air gases.

Eight Rules For Successful Smoke Ejection By Means of Mechanical Horizontal Ventilation

1. Place exhausting smoke ejector in outside opening nearest source of smoke. Since the basic goal is to develop artificial circulation and pull smoke out fast, the ejector should be placed to exhaust in the same direction as the natural wind. Several smoke ejectors may be used simultaneously to create advantageous circulation by placing units so they push or blow products of combustion to exhausting units positioned in outside openings. The objective is to eliminate smoke, and the shortest path is usually the best.
2. During the early stage of fire, most of the heat, smoke and congested fumes rise and accumulate near the ceiling. Exhausting smoke ejectors should be placed high for maximum effectiveness in clearing smoke and providing visibility. **Cool fresh air will come in by itself when the heat and smoke are removed.**
3. Check the direction of the wind and use it to push or blow the products of combustion into an exhausting smoke ejector. When using both exhausting and blowing ejectors, first clear the area from the windward side in the direction of the exhausting smoke ejector. Then move the blowing smoke ejectors toward the exhausting ejector, keeping the circulation line as straight as possible. The venturi action of the straight line circulation will suck most of the smoke from corners.

Material on pp. 96-113 reprinted with slight editing by permission of Super Vacuum Manufacturing Co., Inc.

If a smoke ejector must be directed into the wind, angle it so expelled contaminants cannot reenter the building through nearby openings.

4. Air follows the path of least resistance. If air is allowed to recirculate around the sides of the smoke ejector and in and out of nearby openings, it causes a churning action that reduces efficiency. When a fan is placed in a doorway, it creates a partial vacuum in the room. If the area surrounding the fan is left open, atmospheric pressure pushes the air through the bottom of the doorway and pulls the smoke back into the room.

   Prevent churning

   To prevent churning air, cover the area around the unit with salvage covers or other material, and fasten in place with a spring-loaded rod or similar device. Ideally, the entire opening, with the exception of the area occupied by the smoke ejector, should be covered to attain maximum operating efficiency.

5. Establish desired draft path and keep the airflow in as straight a line as possible. Every corner causes turbulence and decreases efficiency. Avoid opening windows or doors near the exhausting smoke ejector unless opening them definitely increases circulation.

6. Remove all obstacles to the airflow. Even a window screen will cut effective exhaust by half. Avoid blockage of the intake side of the smoke ejector by debris, curtains, drapes or anything that can decrease the amount of intake air. Do not allow the smoke ejector to blow directly against a wall or door.

7. Have smoke ejectors ready to use when hoselines are in position, and have a charged line at hand when actually placing the ejectors. Since smoke is composed of heated gases, it rapidly condenses and settles on everything in the room or building. Therefore, ejectors should be put into operation without delay; however, be sure they are positioned so they do not blow into the building. **They must pull the smoke out.**

   Have charged lines ready

8. When the smoke begins to dissipate, speed up the clearing action with a blowing smoke ejector. Place an exhausting smoke ejector on the lee side of the room or building, where it can pull out the remaining smoke and fumes. Then put the blowing ejector in an outside opening on the windward side, where it will admit fresh, clean air. Results are usually most satisfactory when the blowing ejector is located in the lower part of the window or door.

   In residences and smaller buildings use several blowing ejectors at outside openings or suitable windows in the lowest floor or basement. Close all openings on the lower floor and cover the area around the ejectors to create a "pressure-type" circulation to force smoke or odors out of air conditioning or heating ducts and other concealed spaces. An exhausting ejector on the floor above will carry off disturbed smoke.

A blowing ejector at the bottom of a closet will blow in fresh, clean air and force out smoke and odors through the top of the door opening.

PLACEMENT OF SMOKE EJECTORS

Familiarity with various ways of placing smoke ejectors is required if complete efficiency of operation is to be achieved. The placements detailed below cover the most common situations and illustrate proper positioning of smoke ejectors.

Smoke ejectors function most efficiently when they are positioned several feet inside a door. This creates a venturi effect in the doorway. The following guidelines should be used only as an aid in determining the correct placement of ejectors, based on the sizes of the door and the ejector.

- 16-inch ejector exhausting through a 36-inch door should be positioned 6 to 8 feet from opening (Figure 6.8).
- 20-inch ejector exhausting through a 48-inch door should be positioned 6 to 8 feet from opening.
- 24-inch ejector exhausting through a 48- to 60-inch door should be positioned 6 to 10 feet from opening.

In large door openings, place ejector slightly farther back or use multiple ejectors (Figure 6.9).

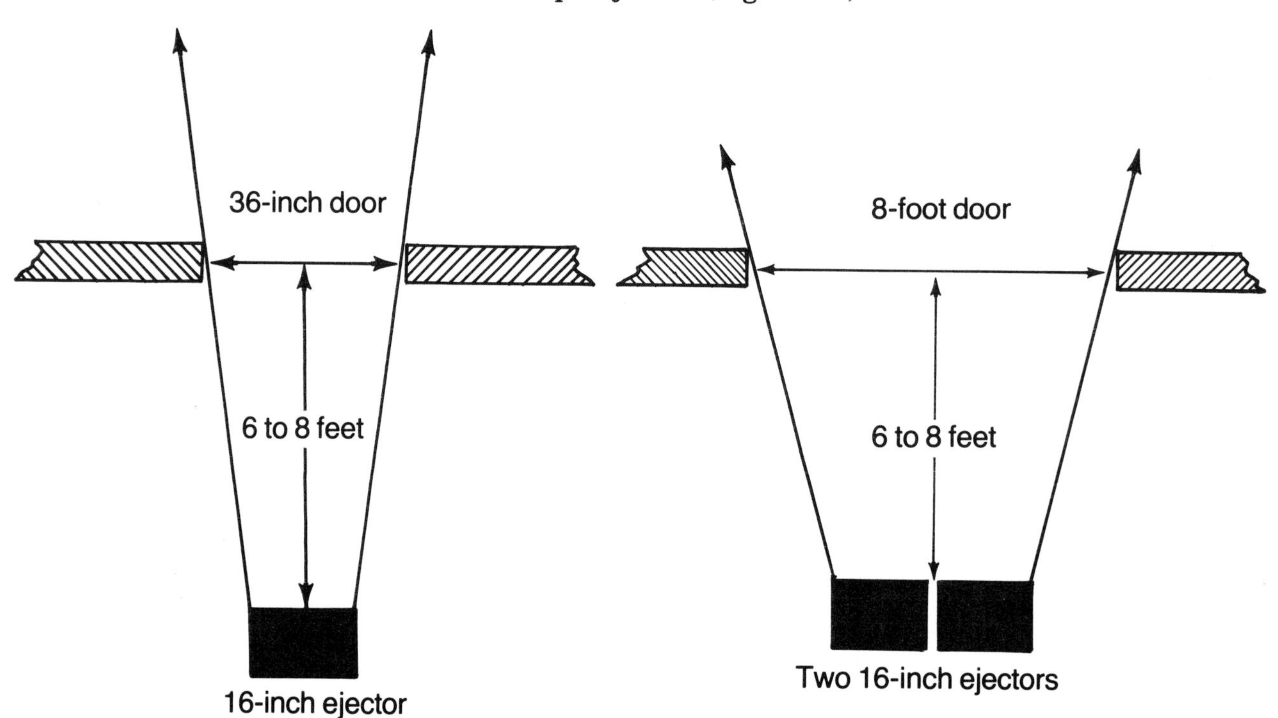

Figure 6.8 — The width of the door and the ejector determine the necessary distance between the two for a good venturi effect.

Figure 6.9 — If the doorway is very wide, more than one ejector may be necessary.

Positioned in this manner the ejector not only moves the air passing through the propeller, but also draws air into the air stream on the discharge side. Efficiency increases by 20 to 100 percent of rated capacity. Actual increase varies according to the size of the smoke ejector, the size of the door opening through which it is exhausting and the size of the opening providing replacement or make-up air (Figure 6.10).

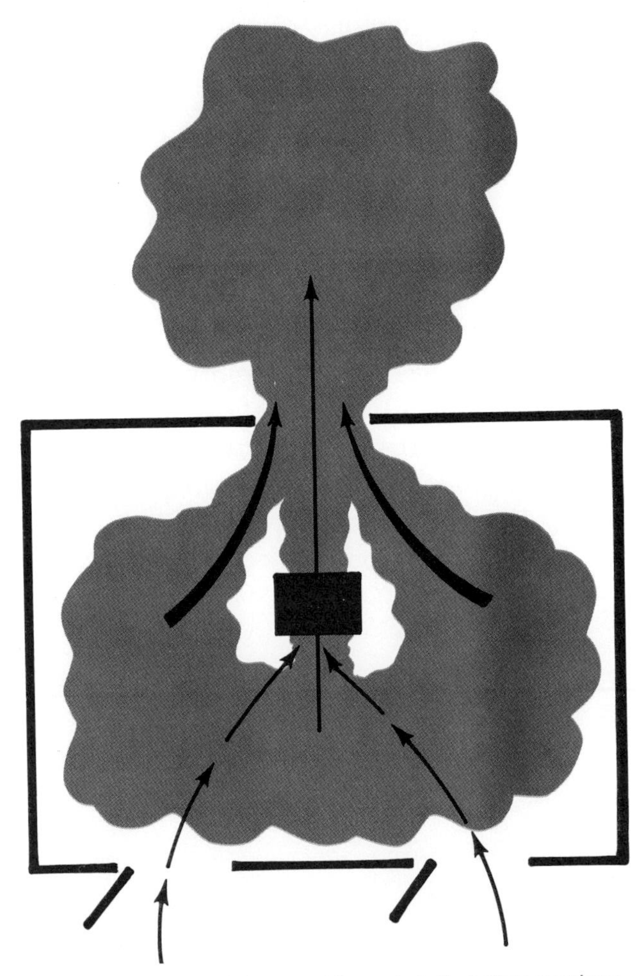

Figure 6.10 — Typical placement for venturi effect. Note openings for replacement air.

The smaller the smoke ejector, the lower the increase in capacity, and vice versa. This is mainly because of the difference in velocity and the concentrated reach of the air stream. By positioning either a 16-inch or a 20-inch ejector 8 to 10 feet inside a 30 by 84-inch door, the volume of air exhausted will increase about the same for either ejector (30 to 40 percent).

Air flow from a 24-inch ejector has a much longer, concentrated reach, which allows positioning far enough back from the door to create a good venturi effect. This increases the distance in the air flow between the smoke ejector and the door into which more air can be induced, and can increase the unit's capacity by 100 percent or more.

Remember, the venturi principle is most effective; however, results can be controlled only if the smoke ejector is positioned inside the area to be cleared and the make-up air channeled to that ejector (Figure 6.11).

To achieve the best venturi effect, smoke ejectors must be placed to push air from inside the building to the outside atmosphere. The venturi can be achieved by placing the smoke ejector outside the building and forcing air into the building and out through openings on the opposite side. If this method is used in a building with numerous partitions or contents, the entire structure must be pressurized to force the smoke through the desired openings. Otherwise air might be forced into undesirable areas such as the attic or adjacent buildings, or through an elevator shaft to other floors. Prevailing winds can also create problems. The slightest misdirection of the wind can interfere with the air flow, thus preventing a good venturi effect.

The volume of air moved with the venturi principle depends on the size of the opening and size of the fan through which the air is exhausted. The larger the opening, the more air exhausted, provided adequate replacement air is entering the building. *Note:* When employing the venturi method of smoke ejection, a heavy accumulation of tar will form on the ejector if it is properly used.

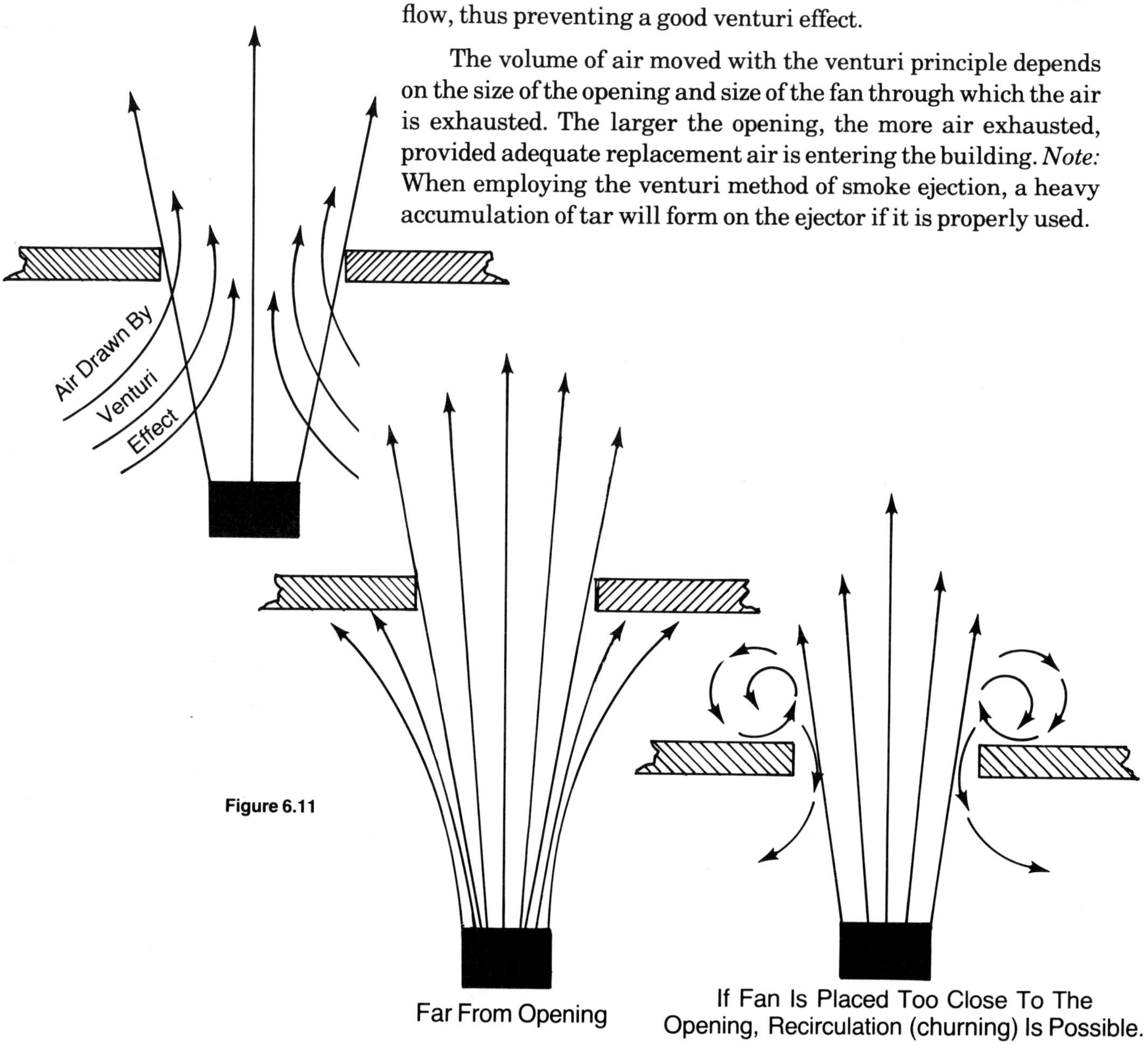

Air Drawn By Venturi Effect

Far From Opening

If Fan Is Placed Too Close To The Opening, Recirculation (churning) Is Possible.

Figure 6.11

THE IMPORTANCE OF REPLACEMENT AIR

Adequate replacement air is of primary importance in achieving good smoke ejection. Inadequate replacement air impairs efficiency of the smoke ejection operation. The opening for replacement air should be equal to or larger than the opening for the exhaust. For example, when exhausting through a common 30 by 84-inch door, the opening for replacement air should also be at least 30 by 84-inches or larger. A combination of windows and other smaller openings may be used to achieve the correct area. Undersized replacement air openings can result in negative pressure, which can overload the smoke ejector motor and cut down efficiency.

Replacement-air opening important

When openings of the desired sizes are not available, it is necessary to increase the amount of air entering the available openings. In many cases another smoke ejector can be placed in the replacement air opening to help pull more air in; however, there are disadvantages in this method. Unless care is used, smoke could be forced into undesirable areas. Also, when air is mechanically blown into a building the turbulence created has a tendency to cool the smoke, causing it to condense and settle faster. Although the smoke dissipates, much of it cools and settles in the building, causing smoke damage.

STANDARD PLACEMENT METHODS

When partitions or other obstructions prevent proper placement of a smoke ejector to create a venturi effect, standard placement methods must be used.

Door Placement

Figure 6.12 shows one of the simplest placements. Attach a hanger to the handles on the unit and hang one of the hooks over the door. Slide the web strap along the handles to change the angle of the air flow or smoke. This method provides air flow through the entire door area and helps eliminate recirculation. By hanging the unit as high as possible on the door, it will pull out the major mass of smoke at the ceiling. Smoke or fumes are expelled quickly, yet traffic can move freely in and out of the doorway. Two units may be hung on a door, if desired.

Hang as high as possible

> An inexpensive hanger for smoke ejectors can be made by bending a concrete reinforcing rod as shown in Figures 6.12d and 6.12e. Scrap pieces of garden hose placed over the hanger protect the door from damage.
>
> There are also several types of commercially available smoke ejector hangers that are spring loaded. The spring inside the unit places tension against the door jamb or the side of the window opening to keep the hanger in place.

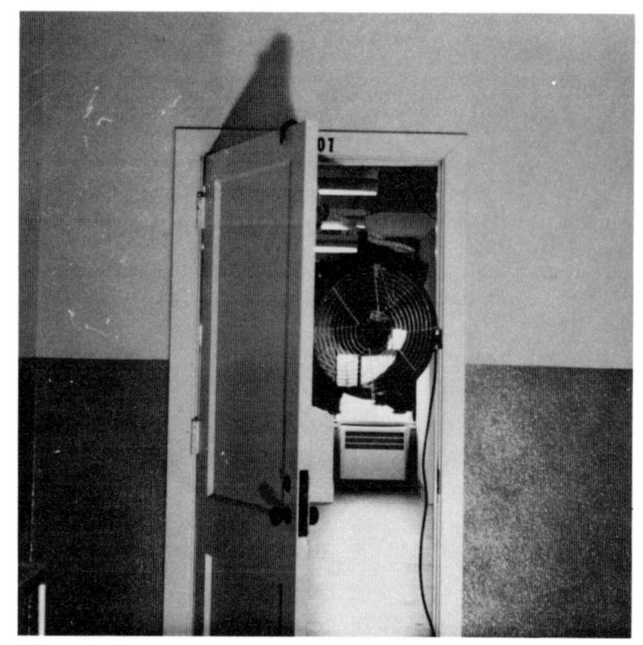

Figure 6.12a

Figure 6.12b

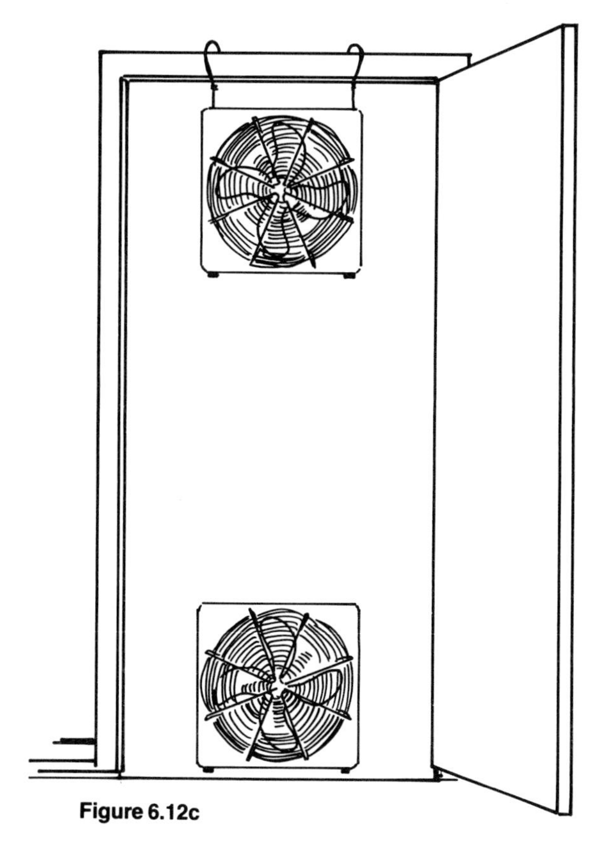

Figure 6.12c

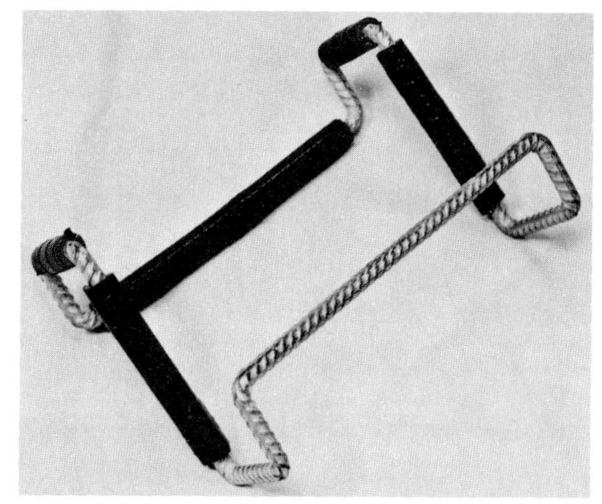

Figure 6.12d

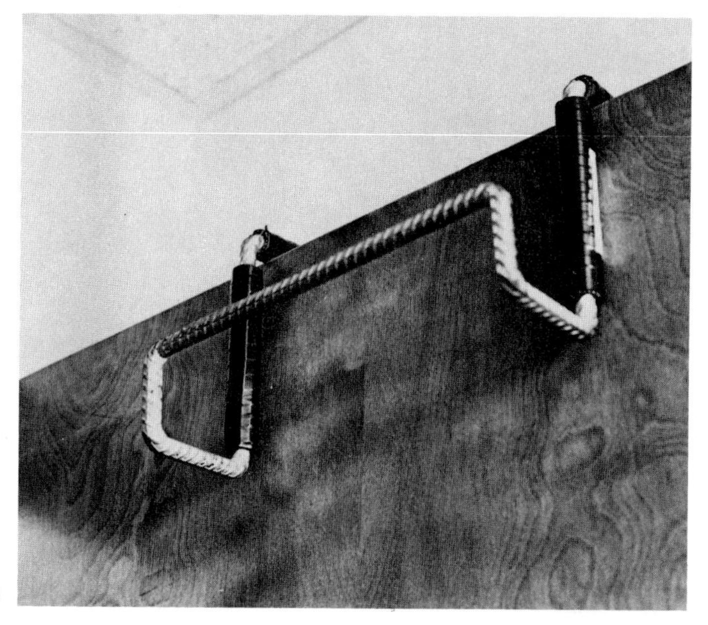

Figure 6.12e

Figure 6.12 — There are several means by which a smoke ejector can be placed in a doorway.

Hook the ejector in a doorway, as high as possible, and tilt unit toward the ceiling to attract the greatest volume of smoke. To change the pulling action to a blowing action, reverse the direction of the smoke ejector. For maximum ejection, cover the area around the unit with salvage covers or other material and fasten in place with a spring-loaded rod or similar device (Figure 6.13). If it is necessary to stir dead air and smoke so it can be pulled from the building by another unit in a different opening, swing the door back and forth to locate the best angle, then keep the door in position with a doorstop. Once connected to an electrical source, the ejector can operate unattended.

Figure 6.13 — Cover the area around the ejector, to prevent churning and to obtain maximum ejection.

Hall or Archway Placement

When the entrance to an area is an open archway, or when doorways are exceptionally wide, place an extension ladder across the opening and extend it until it is wedged tightly between the opposite corners of the opening (an attic ladder might work well for this). The diagonal position of the ladder provides a good support for the smoke ejector and allows room for traffic to pass through the archway. Determine the best direction for the air flow and hang the unit from the top rung of the ladder, as close to the ceiling as possible (Figure 6.14). Other units may be placed along the ladder to create air movement. To change the angle of the unit, slide web straps along the handles. Run the electrical cord down the ladder rungs to keep it safely out of traffic.

Floor-Opening Placement

When smoke, heat or congested fumes are trapped in a windowless area underground, it is necessary to make an opening for their release. Select an opening — such as a door — above grade through which to channel the smoke. An outside door is best. Cut a hole in the floor behind this opening. Place a straight ladder flat

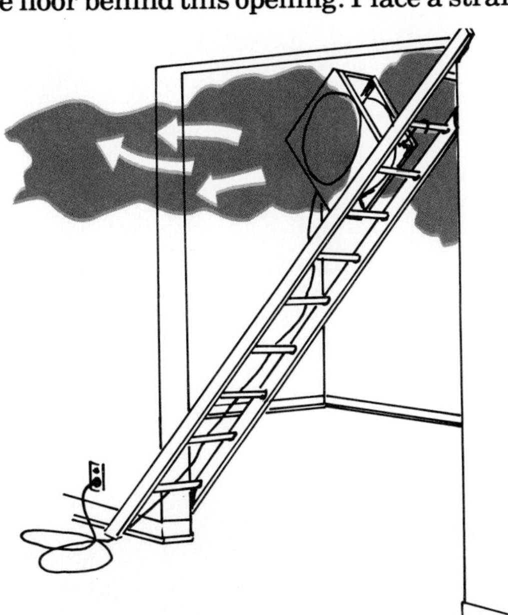

Figure 6.14 — Hall or archway placement.

across the hole in the floor. Use a pike pole to prop another ladder over the first ladder. The two ladders will form a horizontal V with the large end extending through the doorway. Raise the second ladder to the height that gives the best angle and position for smoke ejecting. Cover the top and sides of the ladder framework with a tarp, being very careful to tuck the tarp under the end and sides of the ladder on the floor so it is held in place and away from the hole and the ejector. This construction can also be done away from the hole and then the assembly slid over the hole (Figure 6.15). If there is no outside opening adjacent to the area where the hole must be cut, place additional units so they will force smoke to the closest outside opening.

For additional power, another unit can be hung in the center of the ladder framework, forcing the smoke away from the ladder and tarp. Spray tarp and ladders to prevent scorching. Cover or barricade the hole in the floor when removing equipment.

Stairwell Installation

Place an appropriately sized ladder on stair step, keeping it to one side so traffic can move past. Brace against the wall at a slight angle. Put hanger hooks over the ladder rung at a height that will enable the ejector to pull smoke from the lower ceiling (Figure 6.16).

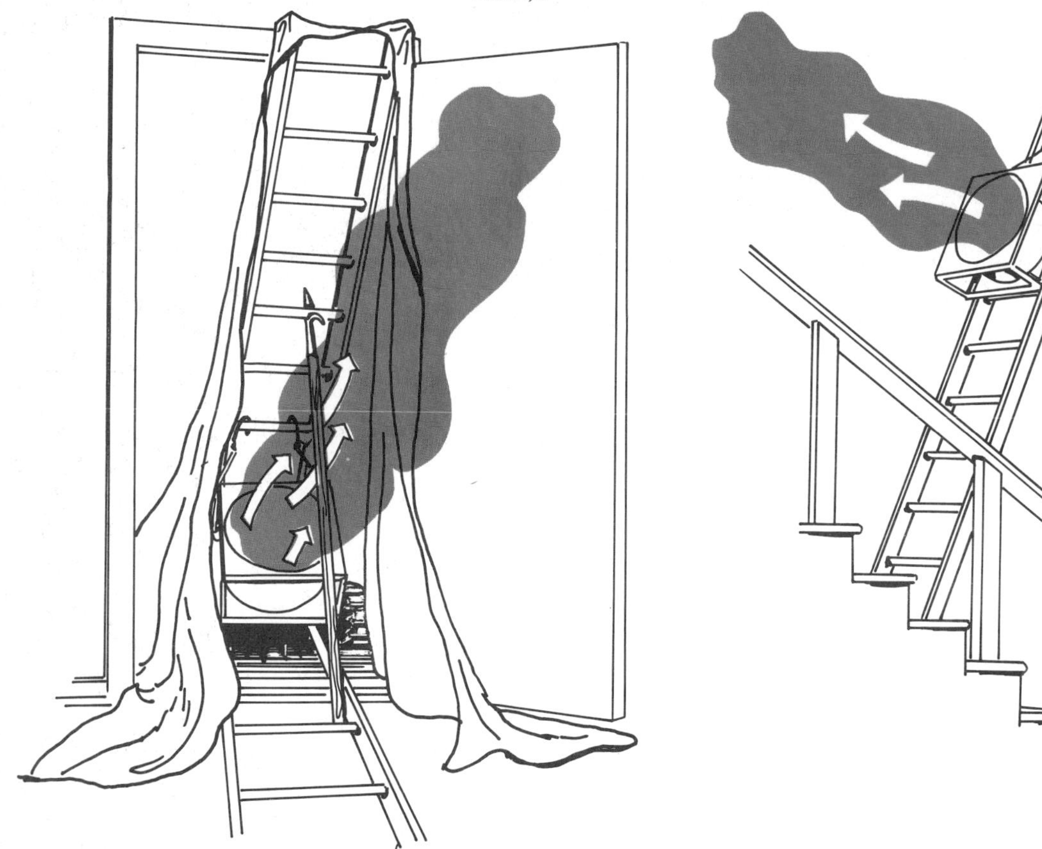

Figure 6.15 — Floor opening placement.

Figure 6.16 — Stairwell placement.

A-Frame Installation or Stepladder Hang

This placement is particularly effective in areas where there are no supports for a hanging-type installation, such as long hallways or large rooms. The A-frame, made from two ladders or a stepladder, provides sufficient support for a complete bank of units. With multiple hangings of units, a huge air mass can be blown in or great quantities of smoke and fumes pulled out. Steady the installation by footing the heels of one of the ladders with a pike pole or piece of lumber lashed across the heels to the bottom rung. After checking to be certain the entire assembly is firmly set, hook ejectors to ladder rungs with hanger units (Figure 6.17). Before plugging in the units check the angles of the units to be sure the air or smoke movement is properly directed. Once the assembly is operating, firefighters can be released for other functions.

Figure 6.17 — A-frame placement or stepladder hang.

Window Placements

There are as many ways to use the unit in window locations as there are types of windows; however, two principal kinds of windows are encountered: the double-hung window and the casement window.

Figure 6.18 — Double-hung window placement.

DOUBLE-HUNG WINDOW

- Raise lower window.
- Place unit on windowsill, aim away from exposures, and pull window down onto handles (Figure 6.18).
- Secure both hooks of the hanger unit into bottom rail of the window.
- Wrap cable of hooks around handle to obtain length that will hold smoke ejector securely on lower rail of window.
- Stuff open areas with pillows or cover with blankets or salvage covers to prevent churning.

Properly installed, the unit performs unattended and ejects smoke at maximum efficiency. *Note:* Be sure the unit is correctly placed. To eject smoke, fan blade must be on the inside forcing air over motor to outside. Reverse the position to pull in fresh air.

CASEMENT WINDOWS

- Roll window open and place one hanger hook over top window hinge.
- Fasten the second hook into the mullion of the window to steady the unit.
- Rest unit on metal edge of window to prevent damage.

This placement is of tremendous value in venting the common mattress fire or similar one-room fires (Figure 6.19).

LADDER PLACEMENT FOR WINDOW

The use of a ladder allows a unit to be placed in a high opening when a more convenient support is not available (Figure 6.20). Ladders can be set up either inside or outside the building. Place the ladder against the wall, determine the best rung for the

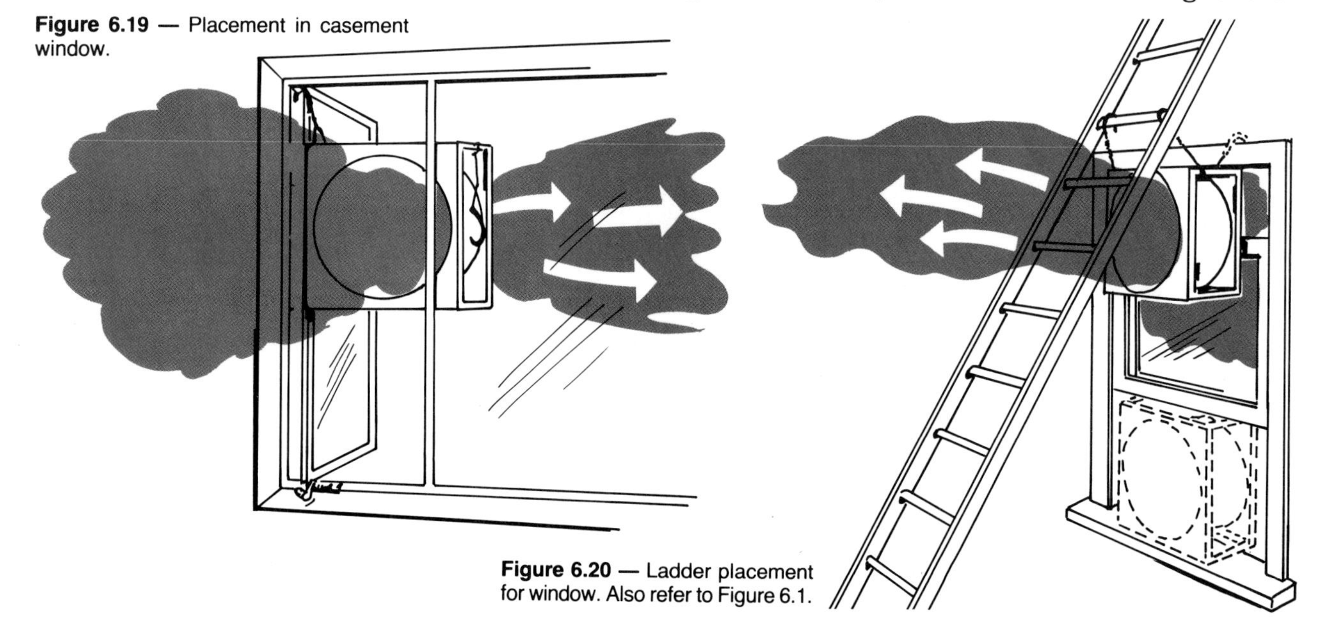

Figure 6.19 — Placement in casement window.

Figure 6.20 — Ladder placement for window. Also refer to Figure 6.1.

ejector and hang the hanger hook over the rung. Once the unit is plugged into an electrical outlet, it can be left alone to perform its job.

WINDOW WELL PLACEMENT

Window well placement is ordinarily set up on the outside to pull smoke and fumes from a basement (Figure 6.21). Since the ejector rests on or near the ground, a larger unit (24-inch) can be used for greater blowing or pulling power.

It is usually necessary to break the glass in the window. Cover the remaining area around the unit with a tarp or similar material to increase pulling power of the unit. The shape and location of the window might require tilting the units in order to get the closest proper angle to the window itself.

Figure 6.21a

Figure 6.21b

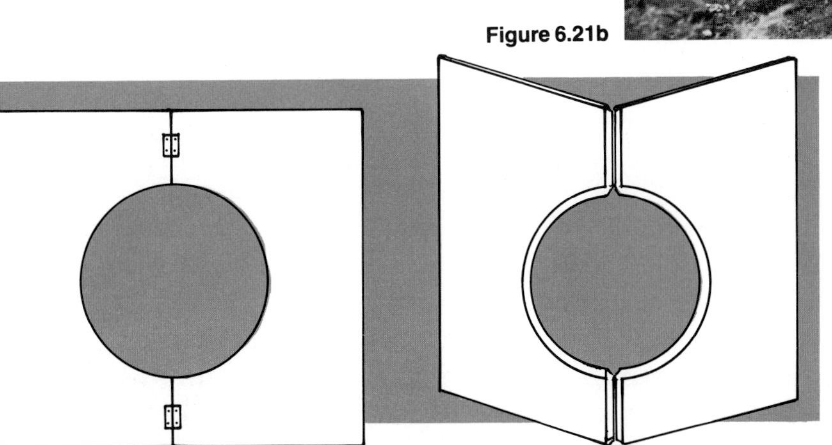

Figure 6.21c

Figure 6.21 — There are several methods by which a smoke ejector can be positioned over a window well. Hinged plywood board shown in Figure 6.21c is easy to construct and store.

SCUTTLE HANGER PLACEMENT

This specialized and highly efficient method of using the unit pulls smoke out of attics, manholes, or similar types of openings. Smoke or fumes concentrated in such a tight area will be compressed to extreme density. The direct and expedient use of the unit in a comparatively small opening effectively ejects this dense smoke.

Suspend the smoke ejector firmly into the scuttle hole opening with a pike pole and hangers (Figure 6.22). If a pole is not immediately available, use the alternative method of suspension, as shown by the dotted lines in the illustration. One person can easily install a smoke ejector in a floor opening or manhole; however, it may be expedient to use two persons to speed up the installation.

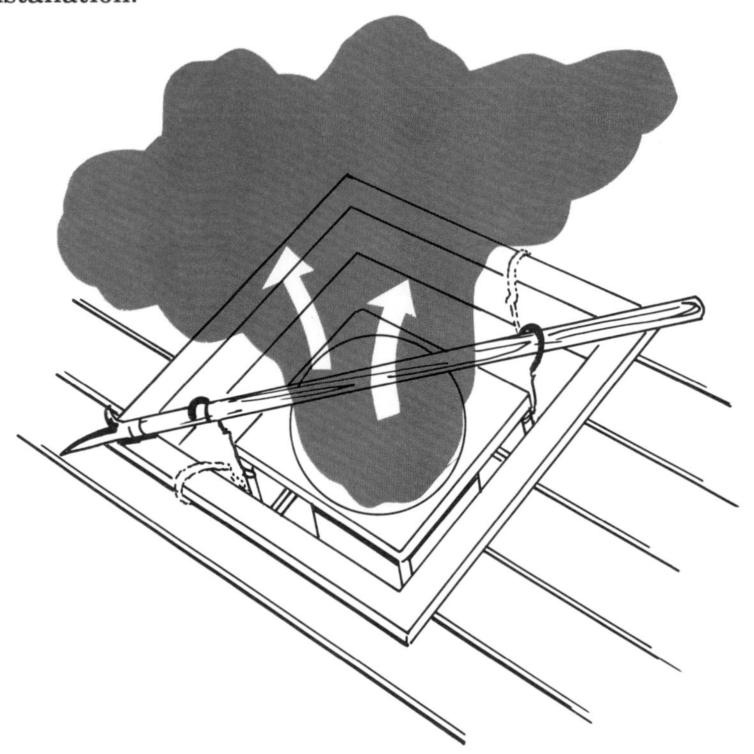

Figure 6.22 — Scuttle hanger placement. Hinged board shown in Figure 6.21c would work well here, too.

HEAVIER-THAN-AIR GASES

Because cooled or unheated heavier-than-air gases concentrate in low points of a building, especially basements, pits and tunnels, they can be removed only by mechanical ventilation. This is accomplished by following the rules for normal mechanical ventilation with the exception of Rule No. 2. Heavy gases accumulate at or near the floor; therefore, the ejector must be placed as low as possible in the opening. Position several ejectors to push gases to an exhausting unit, which in turn will expel gas to the outside atmosphere. When a heavy gas is found below grade, create as forceful a circulation as is possible.

Use both pulling and blowing actions to dilute heavy gas, as much as is possible, when it reaches the atmosphere.

Since many heavy gases are combustible, all sources of ignition inside and outside the building must be eliminated. Always use smoke ejectors with explosion-proof motors.

Protect all firefighters working below grade with positive-pressure self-contained breathing apparatus because this type of gas can asphyxiate.

FLEXIBLE DUCT ATTACHMENT

To solve the problem of supplying fresh air to confined areas, attach a flexible duct to the ejector (Figures 6.23 and 6.24). The flexible duct can be coupled to the intake side, exhaust side or both sides of a smoke ejector.

Figure 6.23a — The flexible duct attachment collapses to an easily managed length. *Courtesy of Super Vacuum Mfg. Co.*

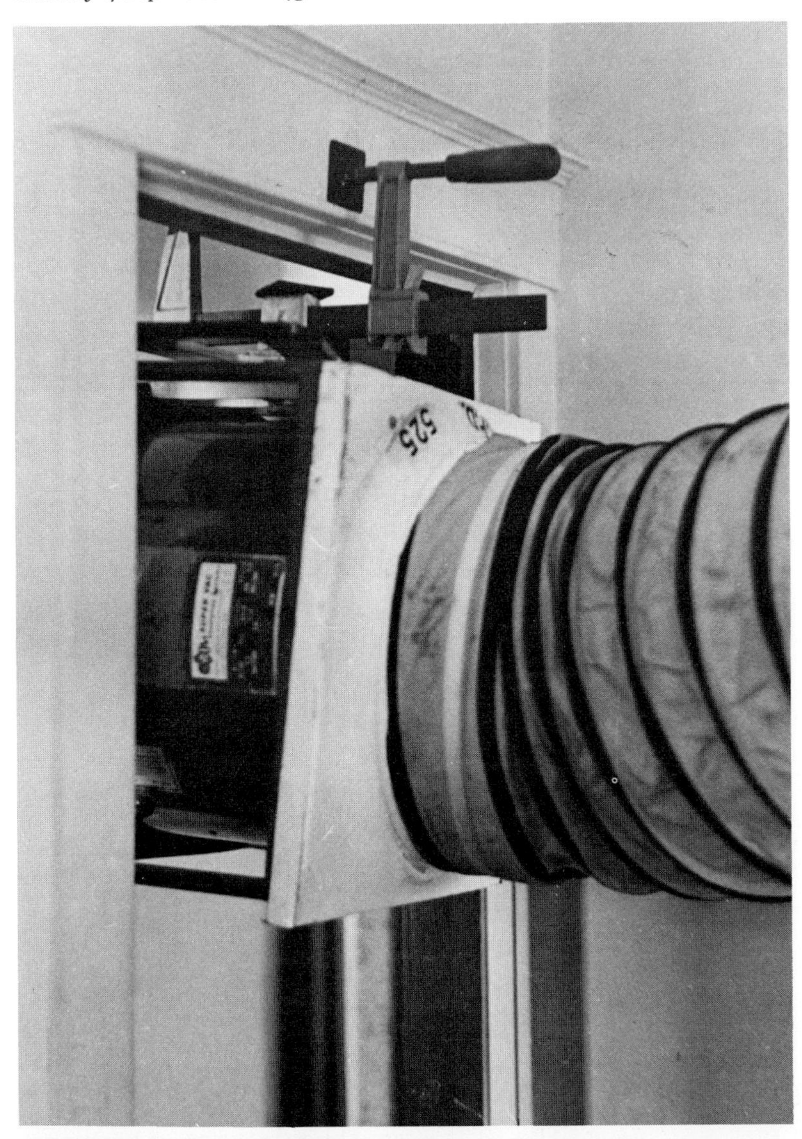

Figure 6.23c — Special hanger attachment that can be used with or without duct. *Courtesy of Super Vacuum Mfg. Co.*

Figure 6.23b — Ducts can be used to supply fresh air to confined spaces. *Courtesy of Super Vacuum Mfg. Co.*

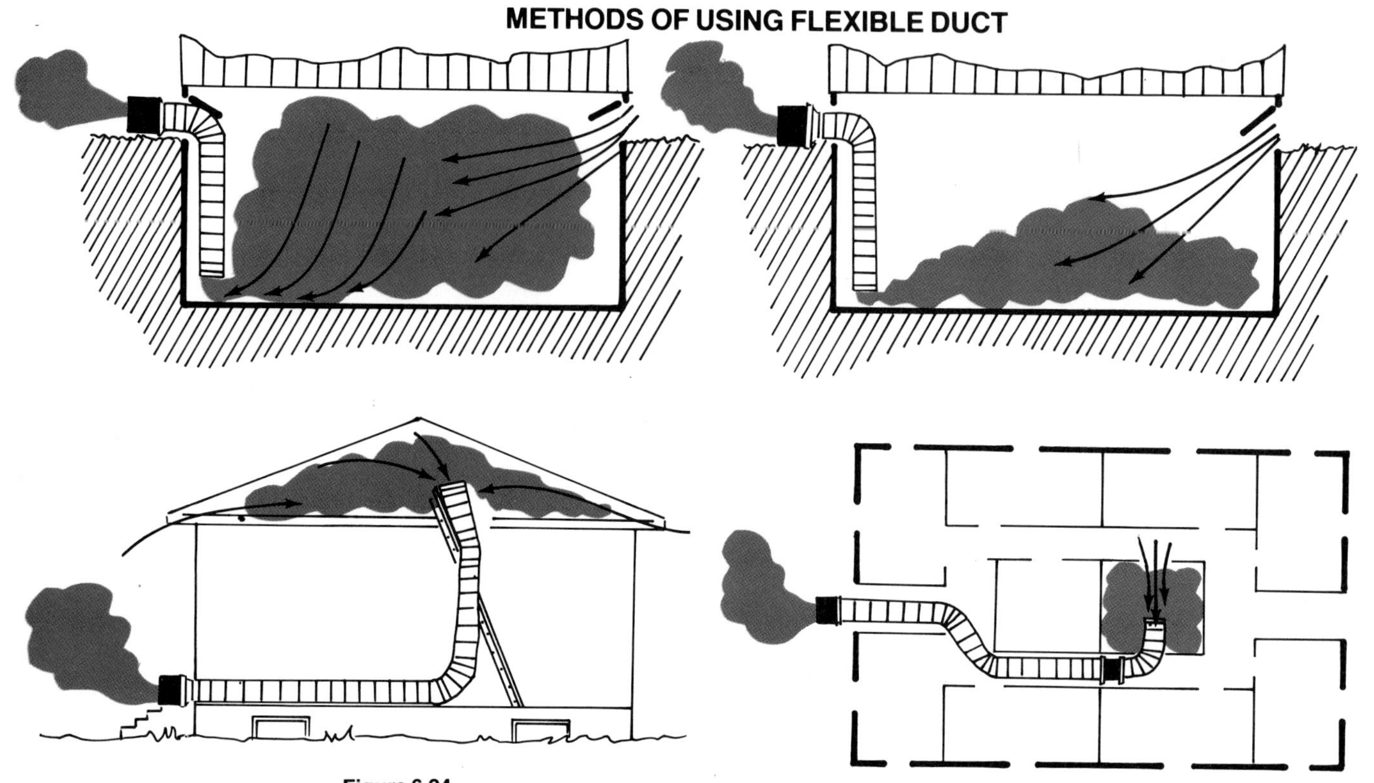

Figure 6.24

To lay a complete temporary exhaust system into any kind of building or confined area, couple two or more ejectors together. This system will ventilate and eject smoke from basements, attics, suspended ceilings, roofs, ship holds, storage bins, railroad cars, and other areas. It provides fresh air for rescue crews in confined spaces such as manholes, sewers, and silos.

Because smoke or fumes channel through the flexible duct to the outside without contaminating other areas, this method is particularly useful in hospitals, schools or shopping malls where there are many internal rooms. Frequently today's modern concrete and steel construction make it extremely costly, if not virtually impossible, to ventilate through the roof. However, with the smoke ejector and flexible duct combination, a temporary exhaust line can be laid to channel the smoke down a hall, through a room, or even through an entire business or house, without causing smoke damage or contamination.

When ventilating with smoke ejectors, it is important to remember that **replacement air must be brought into the area from which smoke is being exhausted.** This can be difficult when ventilating in areas with one door through which the ejector can exhaust but with no opening at the opposite end of the room through which to bring replacement air. However, by placing the smoke ejector in the doorway and laying the flexible duct into the room, it is possible to exhaust the smoke and bring replacement air through the door, around the frame and across the room to the end of the duct.

The ejector/flexible duct combination is also an excellent method for ventilation in areas below grade. It works well for smoke ejection and removal of heavy vapors that settle near the floor and into pockets and low areas. The smoke ejector can be positioned at ground level or above, with the flexible duct running through a window, down a stairway or an elevator shaft into the basement. Replacement air can be channeled through the same opening in which the fan is located or through other available openings.

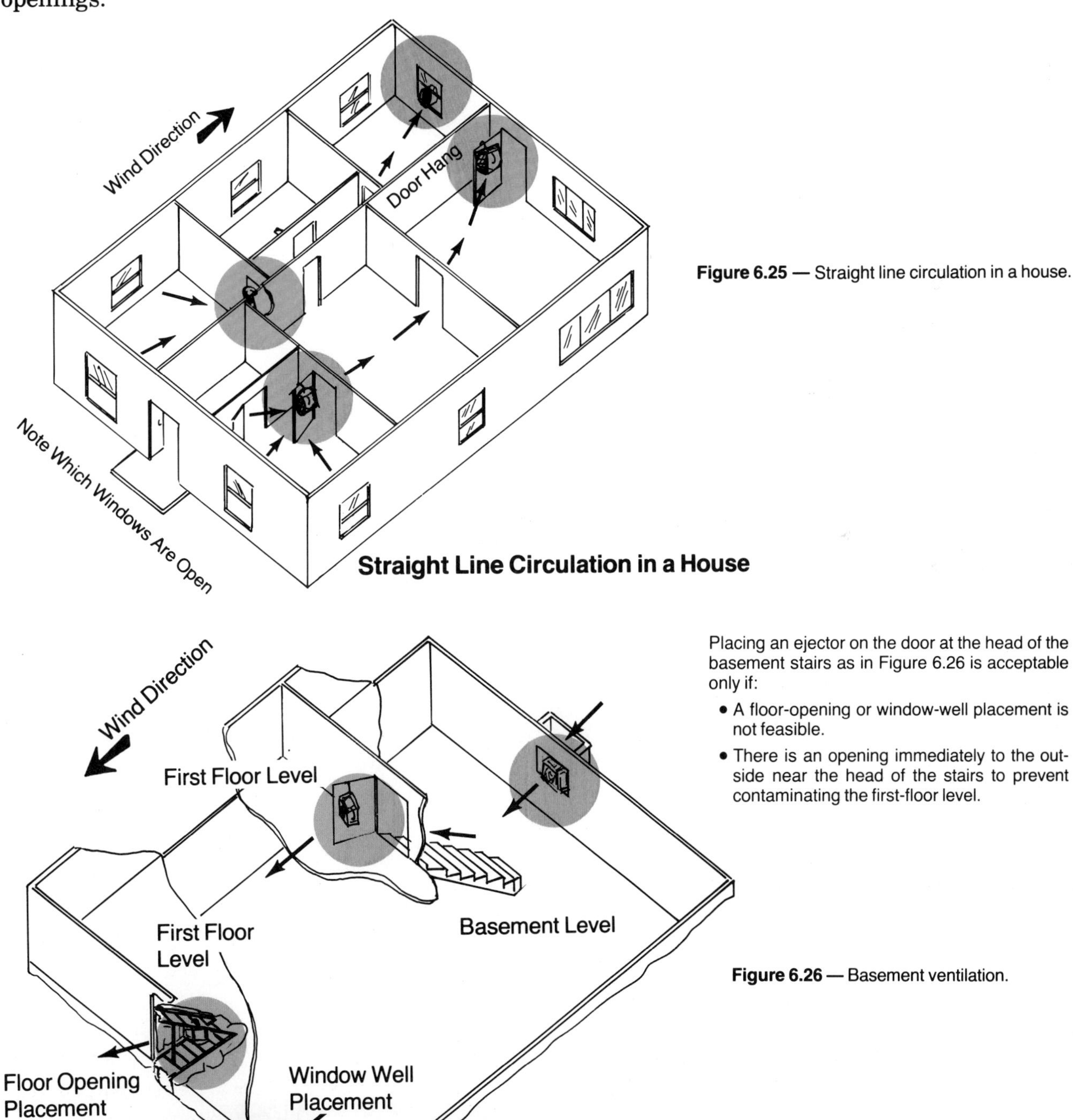

Figure 6.25 — Straight line circulation in a house.

Placing an ejector on the door at the head of the basement stairs as in Figure 6.26 is acceptable only if:

- A floor-opening or window-well placement is not feasible.
- There is an opening immediately to the outside near the head of the stairs to prevent contaminating the first-floor level.

Figure 6.26 — Basement ventilation.

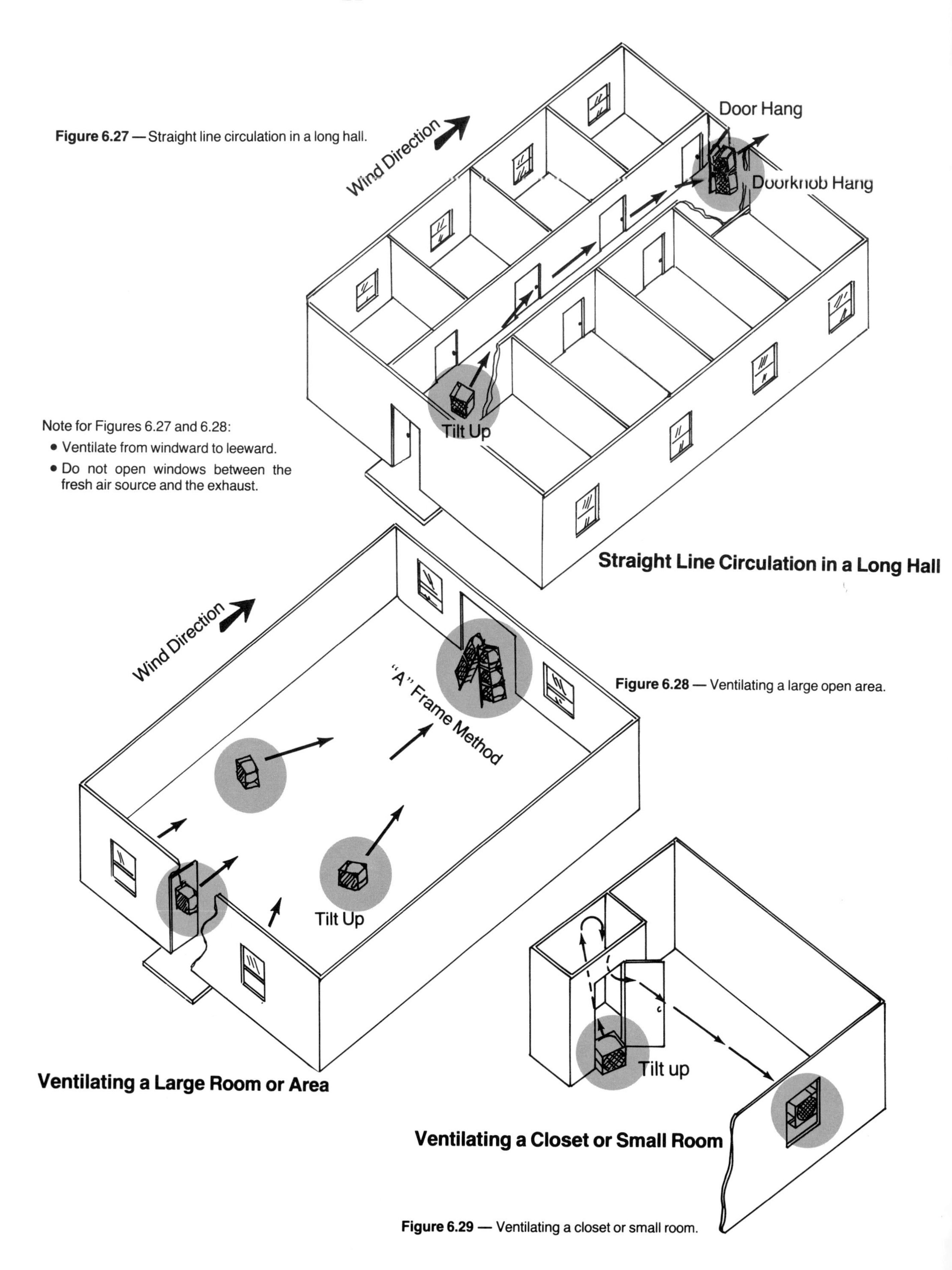

Figure 6.27 — Straight line circulation in a long hall.

Note for Figures 6.27 and 6.28:

- Ventilate from windward to leeward.
- Do not open windows between the fresh air source and the exhaust.

Figure 6.28 — Ventilating a large open area.

Figure 6.29 — Ventilating a closet or small room.

SAFETY PRECAUTIONS

- Pick up the smoke ejector *only* by handles or hangers.
- Prevent curtains and drapes from covering intake of a smoke ejector.
- When blowing smoke ejector is on the floor, supervise traffic in the area. The suction action of the smoke ejector can draw loose clothing into the unit. Keep children away from any operating smoke ejector.
- Do not operate smoke ejectors too long in severely cold weather, especially if the fire has affected the heating system. Prolonged operation can cause water pipes to freeze.

Cold weather operation

- Prevent the blowing about of papers, loose material, debris or dirt, which can add to overall damage or inconvenience.
- Be careful that the increased draft does not close doors and shut off air circulation.
- In combustible atmospheres use smoke ejectors with explosion-proof motors only. Smoke ejectors with explosion-proof motors and switches can be started or stopped safely in combustible atmospheres. Place the smoke ejector in position with the switch in the OFF position. Make all necessary electrical connections (even in the combustible area). Turn the switch ON. Since no circuit is complete with the switch in the OFF position, there can be no arc or ignition of the vapors. When the switch is turned on, the starting arc is contained inside the motor.
- Prevent heavier-than-air gases from entering other low areas where damage or injury could occur. Drive gas forcefully upward into the atmosphere, where it is diluted into a harmless state.

Heavier-than-air gases

- A smoke ejector should never be operated or adjusted while guards are removed from the unit.
- All electrical connections should be grounded.

THE APPLICATION OF WATER FOG AS AN AID TO VENTILATION

Application of water as a fog stream possesses great merit for the control of certain types of fires. The speed with which a given volume of water will absorb heat is in proportion to the amount of water surface that is exposed, and, likewise, its ability to attract or collect the carbon, tar, and ash particles is also increased. As these water particles absorb heat from the heated area, the water is converted into steam, and the expansion that takes place may materially aid ventilation.

When water changes into steam, it expands approximately 1,700 times its volume (Figure 6.30). This means that one cubic foot of water (7.5 gallons) converted into steam expands to fill approximately 1,700 cubic feet of space. It might be well to visualize a nozzle expelling 75 gallons of water fog every minute into an area that has been heated to well above 212° F. During one minute of operation, ten cubic feet of water will have been vaporized, expanding to 17,000 cubic feet of steam. This is enough steam to fill a room approximately 10 feet high, 25 feet wide, and 68 feet long.

In extremely hot atmospheres, steam will further expand to greater volumes. For example, when water is applied into an atmosphere of around 1,000°F and is converted to steam, its relative expansion will be almost double that of water applied into an atmosphere around 300°F. Steam expansion is not gradual, but rapid, and if the room is already full of smoke and gases, the steam generated must force its way to the outside through any available opening, carrying smoke and gases with it. The room will be left hot and humid, but conditions will be such that firefighters can enter with a reasonable degree of comfort.

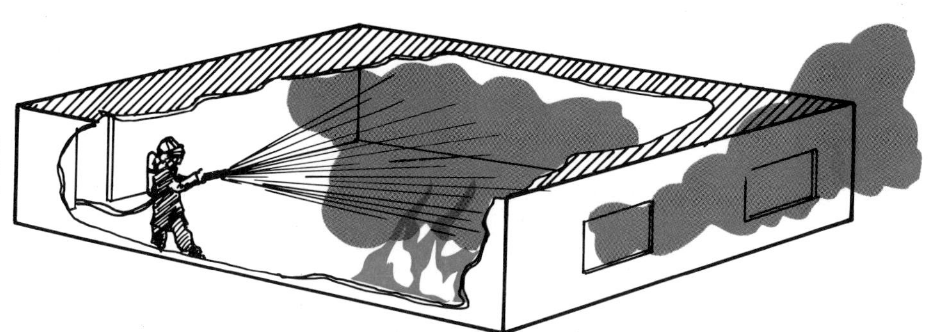

Figure 6.30 — Water converted to steam expands 1,700 times. This expansion absorbs heat and forces hot air and combustion gases out of a confined space.

USING WATER FOG TO EXPEL SMOKE AND GAS

The use of water fog in fire extinguishment and in ventilation requires a special technique of operation. The mere fact that firefighters have a good fog nozzle that supplies a protective curtain does not enable them to advance into a heavily charged area and expect to do an effective ventilating and extinguishing job. When water fog is used for ventilating and extinguishing purposes, the degree of effectiveness depends upon how, where, and when the fog stream is applied.

It has been found that a fog stream directed through a window or door opening will draw large quantities of heat and smoke in the direction in which the stream is pointed.

Compared with mechanical smoke ejectors, fog streams have been found to remove two to four times more smoke, depending on the type and size of the nozzle, the angle of the fog pattern, and the location of the nozzle in relation to the opening of the building. A fog nozzle stream directed through the opening with a 60° angle fog pattern covering 85 to 90 percent of the opening has been found to provide the best results for ventilation (Figure 6.31). The nozzle should be about two feet from the opening. Larger openings permit greater air flow, so a door might sometimes be more beneficial than a window. Whatever the size of the opening, wide angle streams should not be used, because when the water path comes close to a right angle to the air path much of the energy that moves the air is lost.

There are three drawbacks to the use of fog streams in forced ventilation. There will be an increase in the amount of water damage within the structure; there will be a drain on the available water supply, and in climates subject to freezing temperatures there will be an increase in the problem of ice in the area surrounding the building.

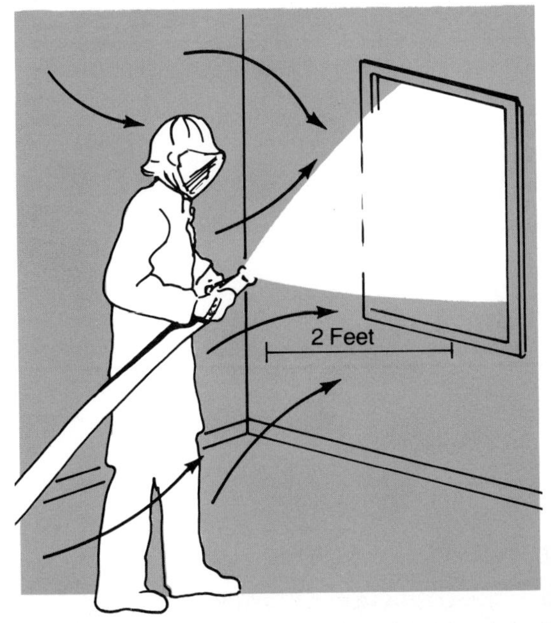

Figure 6.31a and 6.31b — When using a fog stream to ventilate an area, the nozzle should be about two feet from the opening and the pattern should cover 85 to 90 percent of the opening.

MECHANICAL AIDS FOR FOG STREAM VENTILATION

When manpower is limited, fog stream ventilation can still be accomplished with the use of purchased or homemade mechanical aids. One of the commercial nozzle holders is shown in Figure 6.32. A homemade ejector is illustrated in Figures 6.33 and 6.34.

The homemade ejector is made from a 15-gallon drum from which both ends have been removed. A 1½-inch fog nozzle is attached to the inside of the drum with 1½-inch piping. The water supply line is a 1½-inch hoseline. This ejector operates on the venturi principle. The ejector works best at a nozzle pressure of 150 to 200 psi and with a 60-degree fog pattern.

Figure 6.32 — A commercial aid for fog stream ventilation. *Courtesy of Wilcox Silent Nozzleman.*

SMOKE AND HEAT EXTRACTOR PARTS LIST

15-gallon grease barrel

Barrel lid

60 - 80 gpm fog nozzle

1½-inch male hose coupling

1½-inch pipe nipple welded to 1½-inch male coupling

1½-inch PVC female adapter

1½-inch PVC P-trap

1½-inch PVC female adapter

1½-inch pipe nipple welded to female

1½-inch female hose coupling

6 - 8 inch hole cut into center of lid

2½-inch sheet metal collar welded to lid

½-inch strap metal lock ring

6 - 8 inch flexible tubing

1-inch strap metal nozzle support

Strap clamps around male coupling

Figure 6.33a

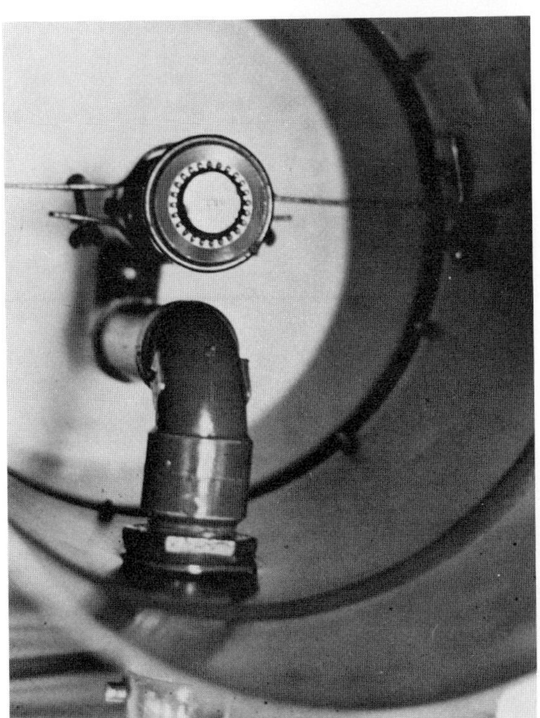

Figure 6.33b

Figure 6.33c

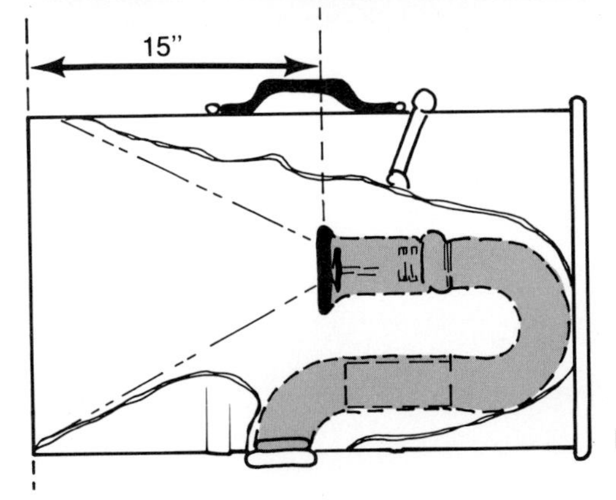

Figure 6.33d

Figure 6.33 — A homemade fog stream ejector can be constructed from readily available parts.

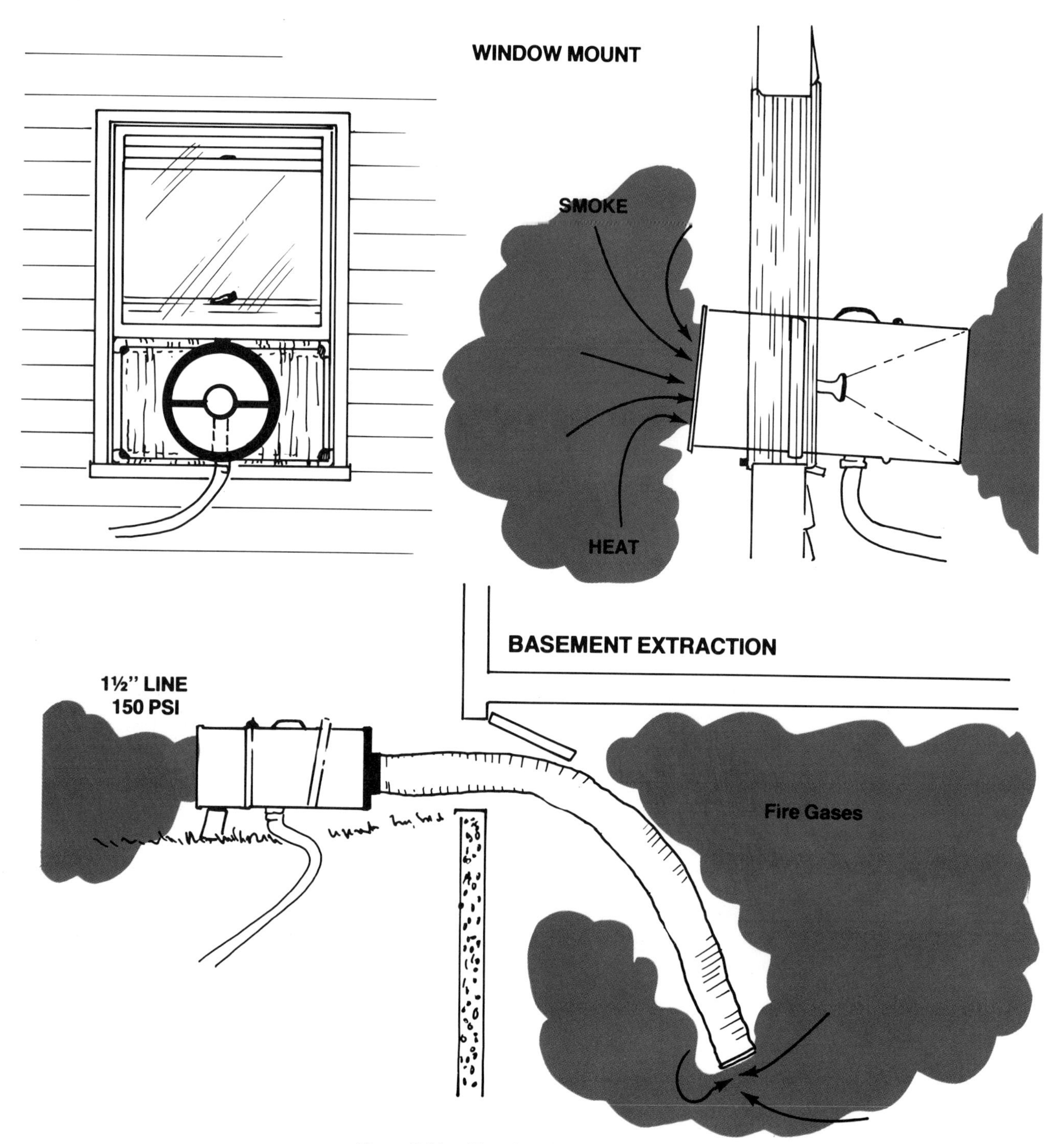

Figure 6.34 — Use of the fog stream smoke ejector is little different from other ejectors. The only difference is that consideration must be given to possible water damage, as well as the route of the smoke.

GLOSSARY

AIR HANDLING SYSTEM The system of fans, ducts, controls, dampers, filters, intakes, exhausts, and heating and cooling equipment installed in a building to circulate air and recondition it.

ARCHED ROOF A roof shaped like an arch.

ASPHYXIATION A condition that causes death because of a deficient amount of oxygen and an excessive amount of carbon monoxide and other toxic gases in the blood.

ATTIC Usually unfinished space between roof and ceiling of the top floor of a building.

BACKDRAFT An "explosion" or rapid burning of heated gases resulting from the introduction of oxygen when air is admitted to a building that is heavily charged by smoke from a fire that has depleted the oxygen content of a building. A backdraft may occur when such a building is opened by the fire department without effecting proper ventilation procedures, or when the fire itself effects an opening. The term *backdraft* probably arose from the fluttering, fluctuating behavior of the smoke immediately before a violent "explosion" of gases involved the structure in flames. The smoke may appear to change direction many times, moving in and out before the explosion. Where heavy smoke is pushing out under pressure, utmost care must be taken in opening up to let out excessive heat without admitting oxygen near the base of the fire.

BALLOON FRAME A building with studs running two or more stories without firestopping.

BOX FRAME A window frame used in masonry buildings. The weight pockets are boxed to keep mortar droppings out while the wall is being constructed.

BREACH An opening made in a wall or other barrier by firefighters to permit rescue, operation of a hoseline, or ventilation.

BREATHING EQUIPMENT Used to protect firefighters from heat and smoke by supplying clean air.

BRIDGING A construction term referring to a structural member that is used to stiffen wood floor joists, wood studding, and steel bar joists used in floor and roof construction. Bridging usually takes the form of cross members, horizontal rods, or solid blocking.

CHARGED LINE A line of hose filled with water under pressure.

CHASE An enclosed shaft containing pipes, ducts, wiring, or other utilities.

CHIMNEY EFFECT A pressure differential caused by gas density differentials resulting from temperature differentials. When the pressure outside a building is higher than inside at a low level and lower at a higher level, air (or smoke) will exit through any opening at the higher level. The chimney effect is also affected by the difference in temperatures between the exterior and interior of a building. When the interior temperature exceeds the exterior temperature, smoke will rise. When the exterior temperature exceeds the interior temperature, smoke will travel to lower floors.

COCKLOFT A concealed space between the ceiling of the top floor of a building and the roof of the structure.

CONDUCTION Method of heat transfer, in which heat is transferred through one substance or object to another substance or object that is touching the first substance or object.

CONFINEMENT Fire fighting operations required to prevent fire from extending to uninvolved areas or to other structures.

CORE PRESSURIZATION SYSTEM A feature of a few of the most modern structures that enables the core area to be subject to separate air handling controls that may, if desired, increase the pressure in the core area during a fire so that heat, smoke, and the products of combustion are kept out of the core area. This is similar to a stair pressurization system.

CRAWL SPACE Area between ground and floor, ceiling and floor, or ceiling and roof, the dimensions of which are such that a person cannot stand up. Often used for ductwork, water pipes, and similar structural adjuncts.

CURTAIN BOARDS Usually found in large-area buildings and constructed of sheet metal or other noncombustible material; extend down from the interior ceiling to delay and limit the horizontal spread of heat by providing the horizontal confinement needed to obtain the desired stack action.

DOME ROOF A hemispherical roof or one formed by a series of rounded arches or vaults on a round or many-sided base.

ENCLOSED STAIRWAYS Stairways enclosed in walls, fitted with fire-resistant, self-closing doors at each floor, continuous from the lowest to the highest floor served.

EXHAUST DUCTS A duct that conveys used or exhaust air from the occupied space.

FORCIBLE ENTRY The process of gaining access to a building by forcing open doors, windows, or any other access point, using various fire department tools.

GAMBREL ROOF A roof with two slopes on each of two sides; the lower steeper than the upper.

HEADER A beam into which floor joists are framed and by which the joists are supported.

HORIZONTAL VENTILATION Ventilation on a horizontal plane; normally cross or forced ventilation.

INCIPIENT FIRE STAGE The first phase of the burning process where the substance being oxidized is producing some heat, but the heat has not spread to other substances in the area. During this phase the oxygen content of the air has not been significantly reduced.

INSULATION A construction material used as a barrier to keep heat from entering, escaping, or flowing between areas of a building. It may vary in form, size, type, method of installation, and degree of combustibility.

JOIST A supporting member used in building construction to support a floor or ceiling.

LEAKAGE The quantity of air, expressed in cubic feet per minute, passing through closed windows, under and around closed doors, and through various building components, and which influences stack effect.

LEEWARD The side of the fire building sheltered from the wind.

MANSARD ROOF A roof with two slopes on each of the four sides, the lower steeper than the upper.

MECHANICAL VENTILATION A process of removing heat, smoke, and hot gases from a fire area by using mechanical equipment such as blowers, exhaust fans, smoke ejectors, and fog streams.

MONITORS The type of vent that normally is rectangular and passes through the roofs of single-story buildings.

MUSHROOMING Term used to describe fire and heat gases spreading out laterally at the top of a structure.

NEUTRAL PRESSURE PLANE The level in a building where the air pressure on the exterior and on the interior are approximately equal and where little gas or air movement or exchange takes place.

NONCONTINUOUS STAIRS Stairs terminating at some point below the roof, thus useless for roof ventilation.

OVERHAUL A process by which firefighters check and search a fire area after extinguishment for any potential smoldering fire, hot embers, etc., so as to be certain that a fire will not start again.

OXIDATION A process by which matter is combined with oxygen.

PARAPET Extension of wall above a roof to prevent fire from extending along the roof.

PARTY WALL A wall common to two buildings.

PENTHOUSES Enclosed structures other than roof structure, located on the roof, extending not more than twelve feet above a roof and used primarily for living or recreational accommodation.

PIPE CHASE/SHAFT See *Chase*.

PITCHED ROOF Any roof elevated in the center, thus forming a pitch to the edges.

PLATE A wood member that supports the ends of roof rafters at the eaves.

PLENUM The void or space above the hanging ceiling when it is used in lieu of registers and ducts for the collection of exhaust, return, or used air prior to its movement via a shaft to the machinery room for reconditioning and recirculation.

POKE-THROUGH Hole provided to draw utility services through a floor. The penetration may negate the fire resistance of the floor.

PRE-FIRE PLANNING Practice of making specific plans for fire fighting operations at designated properties or locations and generally including inspection of the property, classroom instruction regarding procedures to be followed, and preassignment of positions and duties to units assigned to respond.

PRODUCTS OF COMBUSTION Materials put off or released during the burning process.

RADIATION Method of heat transfer by which heat travels through space until it reaches an object.

RAFTER A wood roof joist to which wood roof boarding is nailed.

RESPIRATORY PROTECTION Protection that a firefighter needs to protect the respiratory system from toxic gases and/or dangerous heat. The most common form of respiratory protection is the self-contained breathing apparatus.

RIDGE BOARD/POLE The highest part of a roof. The horizontal member that frames between the upper ends of the roof rafters at the ridge is called the ridge pole.

ROOF DECKING The material directly under the roof covering; usually wood, metal, steel, or concrete.

SALVAGE A process by which firefighters attempt to save property from damage of water, smoke, and heat by removing property from a fire area or by doing anything possible, such as covering property to prevent water damage in a fire area.

SCUTTLES Openings in the ceilings or roofs of buildings, fitted with removable covers for the purpose of providing access and ventilation to the cockloft or roof.

SHEATHING The structural component behind the finished wall, either interior or exterior.

SIZE-UP The mental evaluation made by the officer in charge enabling the determination of a course of action, and includes such factors as time, location, nature of occupancy, life hazard, exposures, property involved, nature and extent of fire, weather, and fire fighting facilities.

SKYLIGHTS Glazed openings, often removable, on roofs over stairways and other vertical shafts that extend to the roof.

SMOKE The visible products of combustion, which vary in color and density depending on the types of material burning and the amount of oxygen present.

SMOKE EJECTOR A gasoline or electrically driven blower device used to eject smoke from burning buildings and sometimes used to blow fresh air into a building to assist in purging smoke or other contaminants.

SMOKE SHAFT A vertical artery, fireproof, and powered by a fan at the top termination, installed in some of the newer buildings for the purpose of removing smoke directly to the outer air from any of the floors served that may become involved in fire.

SMOLDERING PHASE The third stage of the burning process, in which flame may cease to exist if the area of confinement is airtight and lacks enough oxygen.

SOLE OR SILL A horizontal wooden member that rests upon the top of a foundation wall of a building. The vertical framing of the exterior walls and the first-floor wooden floor joists are supported by these members.

STACK ACTION OR STACK EFFECT Also referred to as "chimney effect," the air or smoke movement through a building. Simply stated, cool air enters the lower levels of a building, and warm air within the building rises to the upper levels.

STAIR PRESSURIZATION SYSTEM A system that enables the stairwell to be subjected to separate air-handling controls that may, if desired, increase the air pressure in the stairway so the smoke, heat, and other products of combustion on the fire floor will be restrained therein.

STEAM CONVERSION Water applied to areas of high heat concentration (above 1,000°F) turns to steam rapidly. This water absorbs a tremendous amount of Btus and the generated steam forces smoke and fire gases from the confined fire area.

STRATIFICATION The settling at various vertical levels of accumulations or layers of smoke according to density of weight, the heaviest on the bottom.

STRIP VENTILATION See *Trenching*.

STUD A vertical member used in partitions and walls. Boarding and lath are nailed or fastened to the edges of studs to form a partition or wall. These upright wooden members are usually 2 by 4-inch boards spaced at 16- or 18- or 24-inch intervals.

THERMAL COLUMN Column of smoke and gases, given off by fires, moving upward because heated gases expand and become lighter and rise while cool air bringing additional oxygen is drawn in toward the base of a fire.

TRENCHING Strip or trench ventilation is the process of opening a roof area the width of the building, with a 2-foot wide opening, to channel fire and heat.

TRUSS A construction member used to form roof framework; to support or strengthen as a beam. A truss may form triangles or combinations of triangles.

UNIT VENTS Normally constructed of metal frames and walls and operated by a hinged damper that is controlled either manually or automatically.

VENEER WALLS Essentially frame walls in which the supporting members are wood or metal with a veneer of one layer of brick or stone on the exterior to give the appearance of a solid brick or stone wall.

VENTILATION Applied to fire fighting, the planned and systematic release and removal of heated air, smoke, and gases from a structure, and the replacement of these products of combustion with a supply of cooler air.

VERTICAL VENTILATION Ventilation upward. Also known as top ventilation.

WINDWARD Side of the building the wind is striking.

THE METRIC SYSTEM OF MEASUREMENT

Since the development of the metric system in the eighteenth century the vast majority of the world's countries have converted to the metric system. In fact, by 1974 only six countries were uncommitted to the system.

There are two basic reasons for this conversion. The first is simplicity and efficiency. The metric system is based upon simple powers of ten, where mass, volume and length are all interrelated. Using the power of ten makes computations simple, with a reduction in errors and a saving of time. The second reason is fairly obvious: the simple economics of world trade demand a common system of measurement. The majority of countries use metrics; they supply products in metric measurements. These same countries, as consumers, often demand products in metrics. A country that wants to remain economically competitive in the world market must convert to the metric system.

The metric system used worldwide today, Le Systeme International d'Unites (simply called SI), was agreed upon by the metric countries in 1960.

Familiarity and ease with other systems of measurement are the result of a lifetime of experience. The metric system at first glance may seem much more complicated; however, once proficiency is gained in the use of the metric system the user will wonder why the conversion was not made years ago.

Metric equivalents are given in parentheses after English measures in this manual so firefighters who are accustomed to the English system may become familiar with them. The gallon is the U.S. gallon. To help simplify this introduction to metrics a soft approach has been taken using mostly even numbers rather than the exact figure (hard approach).

SI Rules

The metric system does have its rules, which are necessary, especially because of its international character. They are few and easy to remember.

1. The short versions of the units are symbols in their own right, not abbreviations. No periods are placed after them except at the end of a sentence. The symbol for kilogram, for instance, is kg, not "kg.".
2. Symbols are always written in lower case with the exceptions of the first letter of a symbol for a unit named after an individual and the symbols for the three "biggest" prefixes: Tera, Giga, Mega.

   Therefore, the symbol for kilogram is always kg, never "Kg" or "KG." For Pascal the symbol is always Pa, not "pa" or "PA."

3. The symbols are never pluralized. The symbol for both kilogram and kilograms is always kg, never "kgs." Likewise, 32 square meters is written as 32 m^2, not "32 ms^2".
4. Exponents are used to express powers. Therefore, 32 square centimeters is written 32 cm^2, not "32 sq. cm."
5. If the symbol is composed of two letters there is no space between them — they form a single symbol. Millimeter is written mm, not "m m"; and mm^2 means $(mm)^2$, not $m(m)^2$.

In the lists below, the "soft" metric equivalents are given in parentheses after the "hard" (exact) equivalents.

Length

The basic unit of the metric system is the *meter.* In establishing the length of the meter the developers wanted to use a basic fact of nature. After several changes the meter was finally defined as 1,650,763.73 wavelengths of the orange-red spectographic line of krypton 86. This definition is based on a physical condition that will always be the same, which makes it a true standard. From the basic unit of the meter the system of lengths is simply expanded in either direction in multiples of tens, which eliminates confusing fractions because all lengths can be expressed by using decimals.

millimeter	mm	0.001	m	1 in. = 25.400 mm (25)
centimeter	cm	0.01	m	
meter	m	1.00	m	1 ft. = 0.3048 m (0.30)
kilometer	km	1,000.00	m	

LENGTHS FREQUENTLY USED IN THIS MANUAL

Inches	Millimeters	
0.75	19.1	(20)
1.00	25.4	(25)
1.375	35.0	(35)
1.5	38.1	(38)
1.75	44.5	(45)
2.0	50.8	(50)
2.5	63.5	(65)
3.0	76.2	(75)
4.0	101.6	(100)
4.5	114.3	(115)
5.0	127.0	(125)
6.0	152.4	(150)

Feet	Meters	
50	15.24	(15)
100	30.48	(30)
200	60.96	(60)

As you will notice, the word *meter* is used in developing all other units of length. The prefixes used are either Greek or Latin words or roots. These same prefixes are used in describing other measurements — again, a simplification. Since metrics is in powers of ten it is extremely simple to make other conversions. For instance, how many millimeters are equal to an inch? Knowing that 2.54 cm equals 1 inch, and knowing that 10 mm equals 1 cm, it is simple to deduce that 1 inch equals 25.4 mm.

Metric Squares

The method for devising square measures is simple for example, begin with a square land plot measuring one meter per side. This is a square meter (m^2). A square this size is also equal to one that measures 10 dm (since to dm = 1 m) and therefore is equal to 10 dm × 10 dm = 100 square decimeters. We can then say that 1 m^2 equals 100 dm^2.

Volume

To measure volume, or capacity, develop a cube 1 meter by 1 meter by 1 meter. Its volume is 1 cubic meter (m^3). This cube is the same size as a cube measuring 10 dm to a side, and therefore has a volume of 10 dm × 10 dm × 10 dm = 1,000 dm^3. Consequently, there are 1,000 dm^3 in 1 m^3. A cubic metric unit is always, 1,000 times larger than the next smallest one.

The basic unit of measure for volume is the liter. The liter is defined as 1 kilogram of pure water at 4°D. The liter is also equal in volume to one cubic decimeter (1 dm^3); hence, the interrelationship of measurement of volume and length. To expand the metric volume, all that is needed is to expand the liter by multiples of 10 in either direction and apply the normal prefixes.

milliliter	ml	0.001	L	
centiliter	cl	0.01	L	
deciliter	dl	0.1	L	
liter	l	1.0	L	1 U.S. gallon = 3.785 L (3.75)
kiloliter	kl	1,000.00	L	1 U.S. quart = 0.946 L (0.95)

Temperature

The metric temperature scale is known as the Celsius scale, named after the person who invented it. The term *centigrade,* although technically correct, is not used because of the possible confusion it could cause in some countries where "centigrade" is used to denote a portion of a right angle.

In the Celsius system the freezing point of water is 0° and the boiling point of water is 100°. The difference is 100°, which fits well with the decimal notation of the metric system.

To convert Fahrenheit to Celsius, or vice versa, either of the formulas below may be used.

°C = 5/9(°F 32) or 0.556(°F−32)

°F = 9/5(°C)+32 or °C(1.8) + 32

Pressure

The SI unit used to measure pressure is the Pascal.

Some common pressure measurements used in the fire service are given below with their Pascal equivalents.

1 pound per square inch equals 68.94.757 Pa;

1 psi = 6.895 kPa.

50 psi	344.7 kPa	(350)
80 psi	551.5 kPa	(550)
100 psi	689.4 kPa	(700)

Friction loss per 100 feet of 2½-inch hose

100 gpm	3 psi	20.68 kPa	(20)
200 gpm	10 psi	68.94 kPa	(70)
300 gpm	21 psi	144.77 kPa	(150)
400 gpm	36 psi	248.15 kPa	(250)
500 gpm	55 psi	379.17 kPa	(375)

Velocity (Rate of Flow)

The SI uses liters per second to indicate rate of flow of water. For convenience, the lists below use liters per minute. Note that the symbol is L/min, to prevent the confusion that might result with l/m.

1 gpm = 3.785 L/min (3.75)

Pumper ratings	750 gpm	2,838.6 L/min (2,800)
	1,000 gpm	3,784.8 L/min (3,800)
	1,250 gpm	4,731.0 L/min (4,700)
	1,500 gpm	5,677.2 L/min (5,700)
	2,000 gpm	7,569.6 L/min (7,500)
Hand-lines	20 gpm	75.6 L/min (75)
	30 gpm	113.4 L/min (110)
	100 gpm	378.6 L/min (380)
	120 gpm	454.2 L/min (450)
	200 gpm	757.2 L/min (750)
	250 gpm	946.2 L/min (950)
Head	0.434 psi/foot	9.81 kPa/m(10)

METRIC EQUIVALENTS

PAGE	ENGLISH		METRIC
	CHAPTER 2		
21	eight to twelve inches	=	200 to 300 mm
21	2 × 4-inch	=	50 × 100 mm
21	16-, 18-or 24-inch	=	400-, 450- or 600 mm
38	4 by 4 feet	=	1.21 by 1.21 m
38	10 by 10 feet	=	3 by 3 m
	CHAPTER 4		
66	50,000 square feet	=	4,645 m²
66	40 × 50 feet	=	12 × 15 m
66	20-foot	=	6 m
69	8 by 8 foot	=	2.4 × 2.4 m
69	4 by 4 foot	=	1.21 by 1.21 m
71	4 inches	=	100 mm
73	12 to 18 feet	=	3.6 to 5.4 m
83	25°F	=	-3.89°C
83	250 feet	=	76 m
85	1/8 to 1 inch	=	3 to 25 mm
85	1/8 to 1/2 inch	=	3 to 13 mm
	CHAPTER 6		
98	36-inch	=	900 mm
98	6 to 8 feet	=	1.8 to 2.4 m
98	20-inch	=	510 mm
98	24-inch	=	610 mm
98	6 to 10 feet	=	1.8 to 3 m
98	16-inch	=	400 mm
99	16-inch	=	400 mm
99	24-inch	=	610 mm
99	8 to 10 feet	=	2.4 to 3 m
99	30 by 84-inch	=	770 by 2100 mm
101	30 by 84-inch	=	770 by 2100 mm
114	7.5 gallons	=	28 L

PAGE	ENGLISH		METRIC
114	1,700 cubic feet	=	40 m^3
114	212°F	=	100°C
114	17,000 cubic feet	=	400 m^3
114	10 feet	=	3 m
114	25 feet	=	7.6 m
114	68 feet	=	20.7 m
114	1,000°F	=	538°C
114	300°F	=	150°C
115	15-gallon	=	56-liter
115	1½-inch	=	38 mm
115	150 to 200 psi	=	1,000 to 1,350 kPa
116	60-80 gpm	=	227-302 L/min
116	1½-inch	=	38 mm
116	6-8-inch	=	150-200 mm
116	2½-inch	=	65 mm
116	½-inch	=	13 mm
116	1-inch	=	25 mm

INDEX

IFSTA MATERIALS

FORCIBLE ENTRY, ROPE AND PORTABLE EXTINGUISHER PRACTICES
Types of forcible entry tools and general building construction; use of tools in opening doors, windows, roofs, floors, walls, partitions and ceilings; types, uses, and care of ropes, knots and portable fire extinguishers.

FIRE SERVICE GROUND LADDER PRACTICES
Various terms applied to ladders; types, construction, maintenance, and testing of fire service ground ladders; detailed information on handling ground ladders and special tasks related to them.

FIRE HOSE PRACTICES
Construction, care, and testing of hose and various fire hose accessories; preparation and manipulation of hose for rolls, folds, connections, carries, drags, and special operations; loads and layouts for fire hose.

SALVAGE AND OVERHAUL PRACTICES
Planning and preparing for salvage operations, care and preparation of equipment, methods of spreading and folding salvage covers, most effective way to handle water runoff, value of proper overhaul and equipment needed, determining cause of fire, and recognizing and preserving arson evidence.

FIRE STREAM PRACTICES
Characteristics, requirements and principles of fire streams; developing, computing, and applying various types of streams to operational situations; formulas for application of hydraulics; actions and reactions created by applying streams under different circumstances.

FIRE APPARATUS PRACTICES
Various types of fire apparatus classified by functions; driving and operating apparatus including pumpers, aerial ladders, and elevating platforms; maintenance and testing of apparatus.

FIRE VENTILATION PRACTICES
Objectives and advantages of ventilation; requirements for burning, flammable liquid characteristics and products of combustion; phases of burning, backdrafts, and the transmission of heat; construction features to be considered; the ventilation process including evaluating and size-up is discussed in length.

FIRE SERVICE RESCUE PRACTICES
IFSTA's new rescue manual has been enlarged and brought up to date. Sections include water and ice rescue, trenching, cave rescue, rigging, search-and-rescue techniques for inside structures and outside, and taking command at an incident. The book covers all the information called for by the rescue sections of NFPA 1001 for Fire Fighter I, II, and III, and is profusely illustrated.

FIRE SERVICE FIRST AID PRACTICES
Completely revised. Brief explanations of the nervous, skeletal, muscular, abdominal, digestive, and genitourinary systems; injuries and treatment relating to each system; bleeding control and bandaging; artificial respiration, external cardiac compression and cardiopulmonary resuscitation; shock, poisoning, and emergencies caused by heat and cold; fractures, sprains, and dislocations; emergency childbirth; short-distance transfer of patients and ambulances; conducting a primary and secondary survey.

FIRE PREVENTION AND INSPECTION PRACTICES
Fire prevention bureau and inspecting agencies; fire hazards and causes; prevention and inspection techniques; building construction, occupancy and fire load; special-purpose inspections; inspection forms and checklists along with reference sources; maps and symbols; records and reports.

ESSENTIALS OF FIRE FIGHTING
This new manual was prepared to meet the objectives set forth in levels I and II of NFPA, *Fire Fighter Professional Qualifications, 1974.* Included in the manual are the basics of: hose, ladders, fire streams, forcible entry, rescue, salvage and overhaul, ventilation, fire behavior and science, ropes and knots, extinguishers, protective breathing, fire prevention, fire cause identification, ground cover, communications, water supplies and sprinkler systems.

SELF-INSTRUCTION FOR ESSENTIALS
Self-Instruction book for *Essentials of Fire Fighting.* Covers the most important points of *Essentials,* including NFPA 1001 Firefighter I and II requirements. Perfect supplement for formal training.

FIRE SERVICE PRACTICES FOR VOLUNTEER FIRE DEPARTMENTS
A general, cursory introduction to material covered in detail in Forcible Entry, Ladder, Hose, Salvage and Overhaul, Fire Streams, Apparatus, Ventilation, Rescue and Protective Breathing, and Inspection.

FIRE SERVICE ORIENTATION & INDOCTRINATION
History, traditions, and organization of the fire service; operation of the fire department and responsibilities and duties of firefighters; fire department companies and their functions; glossary of fire service terms.

PHOTOGRAPHY FOR THE FIRE SERVICE
Camera components and operations, films — their advantages and differences, fundamental principles in taking a good picture, processing the negative and print, controlling light for optimum results, accountability and storage of photographic materials; fire scene photography, using photographs as training aids, enhancing fire prevention and public relations through photography, how the investigator can use photographs, glossary of photographic terms.

WATER SUPPLIES FOR FIRE PROTECTION
Importance, basic components, adequacy, reliability, and carrying capacity of water systems; specifications, installation, maintenance and distribution of fire hydrants; flow requirements, flow tests and control valves; sprinkler and standpipe systems.

AIRCRAFT FIRE PROTECTION AND RESCUE PROCEDURES
Aircraft types, engines, and systems, conventional and specialized fire fighting apparatus, tools, clothing, extinguishing agents, dangerous materials, communications, pre-fire planning, and airfield operations.

GROUND COVER FIRE FIGHTING PRACTICES
Ground cover fire apparatus, equipment, extinguishing agents, and fireground safety; organization and planning for ground cover fire; authority, jurisdiction, and mutual aid, techniques and procedures used for combating ground cover fire.

FIREFIGHTER SAFETY
Basic concepts and philosophy of accident prevention; essentials of a safety program and training for safety; station house facility safety; hazards enroute and at the emergency scene; personal protective equipment; special hazards, including chemicals, electricity, and radioactive materials; inspection safety; health considerations.

PRIVATE FIRE PROTECTION & DETECTION
Automatic sprinkler systems, special extinguishing systems, standpipes, detection and alarm systems. Includes how to test sprinkler systems for the firefighter to meet NFPA 1001.

THE FIRE DEPARTMENT COMPANY OFFICER
The officer's functions analyzed; decisions, planning, activating, problem solving, delegating authority and supervision; fire fighting activities of the fire officer.

FIRE SERVICE INSTRUCTOR
Characteristics of good instructor; determining training requirements and what to teach; types, principles, and procedures of teaching and learning; training aids and devices; conference leadership.

FIRE PROBLEMS IN HIGH-RISE BUILDINGS
Locating, confining, and extinguishing fires; heat, smoke, fire gases, and life hazards; exposures, water supplies and communications; pre-fire planning, ventilation, salvage and overhaul; smokeproof stairways and problems of building design and maintenance; tactical checklist.

FIREFIGHTER STUDY GUIDE
Correlates IFSTA publications with NFPA Standard 1001 to guide personnel through the requirements for national firefighter certification at levels I, II, III. Also provides a personal logbook for recording test dates, scores and instructor initials.

PUBLIC FIRE EDUCATION
A valuable contribution to your community's fire safety. Includes public fire education planning, target audiences, seasonal fire problems, smoke detectors, working with the media, burn injuries, and resource exchange.

FIRE PROTECTION ADMINISTRATION

A reprint of the Illinois Department of Commerce and Community Affairs publication. A manual for trustees, municipal officials, and fire chiefs of fire districts and small communities. Subjects covered include officials' duties and responsibilities, organization and management, personnel management and training, budgeting and finance, annexation and disconnection.

INSTRUCTOR GUIDE SETS

Available for *Forcible Entry, Ladder, Hose, Salvage and Overhaul, Fire Streams, Apparatus, Ventilation, Rescue, Inspection, Essentials of Fire Fighting, Aircraft, and for the slide program Fire Department Support of Automatic Sprinkler Systems.* Basic lesson plan, tips for instructor, references. Essentials has NFPA Standard 1001 references and pertinent proficiency tests.

SLIDES

Ladder ***Salvage** **Sprinkler**

2-inch by 2-inch slides that can be used in any 35 mm slide projector; supplements to respective manuals and sprinkler guide sets.

*The complete package consists of the slides, instructor's manual, and instructor's guide sets.

Smoke Detectors Can Save Your Life
Matches Aren't For Children
Public Relations for the Fire Service
Public Fire Education Specialist (Slide/Tape)

TRANSPARENCIES

Multicolored overhead transparencies to augment *Essentials of Fire Fighting,* are now available. Since costs and availability vary with different chapters, contact IFSTA Headquarters for details.

MANUAL BINDERS

Polyflex, expandable-pole binders hold 3 - 10 manuals.

GUIDE SHEET BINDERS

Free with purchase of basic 12 guide sets (*Forcible Entry, Ladder, Hose, Salvage and Overhaul, Fire Streams, Apparatus, Ventilation, Rescue and Protective Breathing,* and *Inspection, Aircraft, and Sprinklers).*

WATER FLOW TEST SUMMARY SHEETS

50 summary sheets and instructions on how to use; logarithmic scale to simplify the process of determining the available water in an area.

PERSONNEL RECORD FOLDERS

Personnel record folders should be used by the training officer for each member of the department. Such data as IFSTA training, technical training (seminars), and college work can be recorded in this file, along with other valuable information. Letter size or legal size.

ip to: Date

ME

RGANIZATION PHONE

DDRESS

TY STATE ZIP

Send to
Fire Protection Publications
Oklahoma State University
Stillwater, Oklahoma 74078
(405) 624-5723
Or Contact Your Local Distributor

ORDER FORM

IFSTA MANUALS

WRITE THE NUMBER OF COPIES OF EACH MANUAL NEXT TO ITS TITLE.

	No. of Each
rcible Entry	
dder	
se	
lvage & Overhaul	
e Streams	
paratus	
ntilation	
scue*	
lf-Contained Breathing Apparatus**	
st Aid	
spection	

	No. of Each
Essentials	
Self-Instruction for Essentials	
Volunteer	
Indoctrination	
Photography	
Water Supplies	
Aircraft	
Ground Cover	
Safety	
Private Fire Protection	
Company Officer	

	No. of Each
Instructor	
High-Rise	
Study Guide for Essentials	
Public Fire Education	
Fire Protection Administration	
Manual Binder	
SLIDES	
Ladder	
Salvage	
Sprinkler Systems	

	No. of Each
Public Fire Education Specialist	
Smoke Detectors Can Save Your Life	
Matches Aren't For Children	
Public Relations for the Fire Service	

TRANSPARENCIES
Multicolored overhead transparencies to augment IFSTA 200, *Essentials of Fire Fighting,* are now available. Since costs and availability vary with different chapters, contact IFSTA Headquarters for details.

vailable fall of 1981.

**Available fall of 1981.

OTHER MANUALS AND MATERIALS MAY BE ORDERED BELOW:

UANTITY	TITLE	LIST PRICE	TOTAL

or Free Subscription to Speaking of Fire ☐

btain postage and prices from current IFSTA Catalog or they will be inserted by ustomer Services.

ote: Payment with your order saves you postage and handling charges hen ordering from IFSTA headquarters.

ayment Enclosed ☐ Bill Me Later ☐

SUBTOTAL $

Discount, if applicable $

Postage and Handling, if applicable $

TOTAL $